Internet of Augmented Me, I.AM

Internet of Augmented Me, I.AM

Empowering Innovation for a New Sustainable Future

Patrick Duvaut
Xavier Dalloz
David Menga
François Koehl
Vidal Chriqui
Joerg Brill

iSTE

WILEY

First published 2020 in Great Britain and the United States by ISTE Ltd and John Wiley & Sons, Inc.

ISTE Ltd
27-37 St George's Road
London SW19 4EU
UK

www.iste.co.uk

John Wiley & Sons, Inc.
111 River Street
Hoboken, NJ 07030
USA

www.wiley.com

Library of Congress Control Number: 2020941142

British Library Cataloguing-in-Publication Data
A CIP record for this book is available from the British Library
ISBN 978-1-78630-508-4

MIX
Paper from responsible sources
FSC® C013604

Contents

Forewords

Gary Shapiro, President and CEO, Consumer Technology Association (CTA), USA

Thanks to the rise of tech-friendly leaders and initiatives such as La French Tech, French innovation is making its mark on the world. Each year at CES® – the world's largest, most influential technology event – I'm blown away by the creativity and sheer style of the innovations that French companies exhibit. And I'm increasingly impressed by the steps France has taken to promote a strong tech ecosystem – whether it's permitting companies to test self-driving vehicles on the nation's roads and highways or creating a work visa designed to attract top tech talent from around the globe.

This book exemplifies the growth of French innovation we've seen in recent years. At the center lies what the authors call "the Internet of Augmented Me" – a phrase that summarizes the complex relationship between human choice, digital solutions and environmental factors that will drive our future. I'm excited to see what will happen in the world of French innovation over the course of this decade.

As sister democracies, committed to personal liberty and self-determination, the United States and France share a long and proud history. We must remember this history as the pace of technological and political change continues to increase. Isolationism is not rational in a world where people can share ideas, information and stories with a few swipes, taps and tweets. By recommitting ourselves to our nations' ideals, we can create a global tech ecosystem where everyone – regardless of where they come from or how they identify – can grow and thrive.

The authors bring together a remarkable array of ideas and examples – giving you an image of what the future can and should be. This book is a great way to kick off the 2020s – a decade that, as these pages show, is full of possibilities and opportunities to make our world a safer, stronger and more sustainable place.

André Joffre, President, FNBP (National Federation of Banques Populaires) and Michel Roux, General Director, FNBP, FRANCE

It is a great honor that the authors wanted Banques Populaire to preface this book. Banques Populaire is the fruit of a humanist movement that pushed, at the end of the 19th Century, small artisans and traders to pool their savings to come to the aid of those of them who could not have access to credit, then reserved by traditional banks only for large industry.

The French spirit likes to question established orders and Banques Populaire has been nourished by this audacity of transgression, of a form of protest cooperation, which pushes to innovate and to undertake. It also made the choice of cooperative status to exercise its activity because it highlights relational, geographic proximity and long-term vision. Thus, year after year, it reserves more than 85% of its results, thus creating a new form of mutualization through three means of solidarity: intergenerational, inter-sectoral and inter-territorial.

We find this same spirit in the Internet of Augmented Me (I.AM). It places the human at the heart and it consecrates the advent of the convergence of digital, ecological, environmental, societal, educational, ethical, economic, industrial transitions and health.

I.AM lays the foundation for new industrial, societal and economic models from the Covid-19 era.

The authors also outline the ways for a new era for France, which must be fully aware of its very great capacity:

– to foster new creations of values that increase and protect humans;

– to build the elements of trust essential to the creation of transversalities and allowing joint work;

– to reinvent a lot of our daily lives for a better world.

Creativity and innovation are part of the solutions and France is sitting on a heap of gold.

Thanks go to I.AM for reminding us!

Pascal Faure, Director, INPI (National Institute of Intellectual Property), FRANCE

The world has never changed so fast! Across the planet, the 20th Century saw the birth of a very new way of life, in a context of extraordinary and increasingly globalized innovation in every sector.

France, which was ranked second-most innovative European country in 2019 by the European Patent Office and Europe in general, has substantially contributed to the change, driven by two factors. First, a strong industry in many important fields such as transport, space, energy, telecommunications and chemistry. Second, a tremendous display of skill and success in the new technology sector, especially digital and biotech, proudly vindicated by France today with the massive mobilization of people and resources around what has become the so-called "Start-up Nation" and, more generally speaking, "French Tech".

Such rapid progress on an unprecedented scale has raised new issues of unparalleled complexity. We are at the dawn of a century that holds great promise but also carries enormous risks if we are not able to contain the technological race within a humanist perspective.

We are faced with new expectations – technological, environmental, societal and ethical – and new constraints, which cannot be dealt with separately but must be addressed in an overall approach.

In the forthcoming decades, humanity must take up challenges that are immense and perhaps even existential. Only the structured acquisition and convergence of knowledge will enable humanity to control its future.

Our thanks go to the authors who have shared their reflections on these essential topics. They rightly stress the need to trust in new pathways to adopt the decentralization and collaborative approaches required for our increasingly complex economies. Ultimately, they remind us of an obvious but often hidden truth: intangible assets have become the necessary basis of our evolving world.

Wai Keung Eric Leung, Vice President, Innopark Shenzhen, Founder and COO, Shenzhen XY Interactive Technology, CHINA

The masterpiece I.AM tackles the challenges of redefining growth in the era of innovation and showcasing of the realization of digital value.

The insight gained as a result of the I.AM has underlined the genuine digital value by emphasizing the importance of decentralization, the major role of Blockchain technology, as a whole.

This includes the augmentation of the digital entanglement (the visible inside of the invisible), with I.AM representing the digital "à la carte" guide and the architecture of the future mirror world. Government and industrial leaders, etc. shall act together in an artistic manner with the great predictive intelligence.

Prof. N.K. Goyal, President, CMAI Association of INDIA, Chairman Emeritus, TEMA, INDIA

I am very happy to see this book!

Compliments to the six co-authors, this is a most wonderful, enlightening and updated book on the subject. The digital world has changed the way we live and work. Everything is going towards Digital. Increasingly, it looks like humans will become irrelevant and technology will take over everything in human life. This has brought in related issues of data privacy and security and cyber diplomacy. There is technology greed over human aspirations, relations and respectability.

I am sure that this book will be a revelation for stakeholders and will help in framing appropriate policies for Digital economy in the world.

Introduction

We are at the dawn of a new era, the "Internet of Augmented Me", I.AM.

I.AM catalyzes the "convergence for good" of three worlds – the biological, the physical and the digital, helping us to better tackle the toughest challenges of the 2020s: climate change, resource depletion, an aging population, social inclusion, the empowerment of people and World health crises, such as COVID-19.

People expect to be ethically augmented and cured in a trustworthy manner, while having the final say, giving a genuine sense to their life on a sustainable planet.

As Bill Gates puts it, "Technology is unlocking the innate compassion we have for our fellow human beings."

This is exactly what I.AM deals with. Gary Shapiro, President and CEO of the Consumer Technology Association (CTA) inspired I.AM. The six co-authors are from very diverse backgrounds – industry (Airbus, EDF), start-ups (BTU), consulting (Tasmane), academia (IMT France) and non-profit organizations (Mission CES France) – and they present, in a sharp and comprehensive manner, every angle of I.AM: humans, society, ethics, trust, economics, opportunities, technology, industry and so on.

This book dives into disruptive concepts of I.AM, such as: Trust as a Service, Business as a Game, ATAWAD (AnyTime, AnyWhere, Any Device), Productivity of Collaborative Exchange (PCE), Unimedia, Shazamization of everything, decentralization of everything, BOTization and

Build to Order for Me, Blockchain and Empowerment of Me, edge computing, augmented industry, augmentation value chain and Empowering Innovation, etc.

The fluid, easy-to-read style of this book targets the broadest scope of readers, from purpose-driven and business-oriented individuals, to students, researchers, experts, innovators, consultants, managers and politicians, all eager to empower people to work towards a more sustainable future.

Enjoy reading it!
June 2020

About the Authors

Patrick Duvaut *(Book's initiator and coordinator, Chapter 6)*

Patrick has more than 20 years of international experience (USA, Japan, India, China, and France) as Head of Innovation for large companies, public organizations and start-ups worldwide with core business in telecommunications and digital. Patrick has written four books and more than 200 papers. He owns 70 USPTO patents and three international standards. He is currently involved in frugal, scalable and sustainable blockchain and AI to tackle 21st Century challenges (world health crises such as Covid-19, aging population, social inclusion, people empowerment, resource depletion and climate change). Patrick graduated from École Normale Supérieure, France (MS in Physics and PhD in Statistics) and MIT-USA (Post-Doc in Statistics). He is currently Head of Innovation at IMT France.

Xavier Dalloz *(Chapter 1)*

Xavier has headed the consulting firm Xavier Dalloz Consulting (XDC) since 1993. For more than 30 years, Xavier has performed strategic consulting activities on the use of new technologies in businesses for competitive advantages. Xavier works for some of the largest French companies such as Renault, Air France, Louis Vuitton, L'Oréal, Allianz, La Poste, EDF, ENGIE, BNP Paribas, etc. Since April 1995, Xavier has written a report (40 pages) every month on the same analysis grid. In this report, Xavier analyzes the main issues of "digital" innovations.

David Menga *(Chapter 2)*

David is a research engineer at EDF Lab Paris-Saclay. He is interested in the design of services for smart homes and smart buildings. He is an expert in embedded systems, AI and blockchain. He has been participating in the CES for 16 years and is familiar with the world of high tech. David organizes seminars at the École Polytechnique on the impact of technologies in everyday life. With Professor Nobuo Saito of Keio University, he has co-edited a book entitled *Ecological Design of Smart Home Networks: Technologies, Social Impact and Sustainability*.

François Koehl *(Chapter 3)*

After several years in IT services, François created and developed a strategy and management consulting firm over 17 years before integrating it into one of the large French audit firms, Mazars. After the end of the integration phase, François joined Tasmane in 2019 to focus on "strategies in a digital world". He gives advice to CEOs and CIOs of middle size companies and large groups on their challenges regarding transformation, digitalization, innovations and change of business models.

Vidal Chriqui *(Chapter 4)*

Passionate about Bitcoin since 2011, Vidal is the inventor of BTU Protocol, the first peer-to-peer booking and transaction protocol. BTU Protocol empowers businesses to engage directly with consumers, thus allowing them to take back control of their physical and digital distribution. Vidal is sharing his passion for decentralization and tokenomics as a regular speaker but also as an academic teacher and startup advisor.

Joerg Brill *(Chapter 5)*

Within his role as Head of Plants and AIT Spacecraft at Airbus Defence and Space, Joerg Brill is responsible for the Pre-FALs, aero- and space structures production and MAIT (Manufacturing, Assembly, Integration, and Test) of space systems. In addition, he is in charge of performance and improvement for the overall production for Airbus Defence and Space, spread over 13 sites in five countries. Joerg holds a Master's Degree in Aeronautical Engineering from the Technical University of Munich.

Acknowledgements

The authors are particularly grateful to the following individuals who made very valuable contributions to the book: Vincent Bastien, Ruben Carvajal, Valentin Dillenschneider, François Pistre.

1

With the Internet of Augmented Me, a New Wave of Innovation is Coming, Which Will Change Everything

"We always overestimate the change that will occur in the next two years and underestimate the change that will occur in the next ten. Don't let yourself be lulled into inaction."

Bill Gates

1.1. The Internet of Augmented Me: new technologies, new jobs, new business

The IoT (Internet of Things) universe is exploding. Just look around and see the smart gadgets, clothing, cameras and virtual assistants that populate homes and are carried by people on their travels. The number of IoT devices connected to broadband networks is already significant, but it is far from having reached its maximum.

According to the main forecasts, 22 billion devices will be connected to the Internet by 2025, which is three times their number today.

For companies, the goal is to get a competitive advantage using these objects with the network effect: increasing the number of people or participants who are exposed to the objects or service can increase the value of the company, goods or services.

The explosion of this market has a very positive side as it will provide consumers with new services and commodities. However, it will also bring new business models, as well as a new complexity and security risks, which could prove to be devastating for millions of people who will, increasingly, rely on digital services and networks in their daily lives.

In this context, all connected objects must have their own digital representations, i.e. *digital twins*, with the following properties:

– *unicity of identification of each object*:

- the possibility of identifying each object is a fundamental notion of the IoT. Each object should own and know its identity in order to evolve in the network,

- each object should be a *legal actor* in its own right,

- the assignment of identifiers should be programmable;

– *suitability for its environment and its context of use*:

- an object must be able to report its status and communicate data collected about its environment: temperature, humidity, vibration level, noise or geolocation;

– this new infrastructure should be *proactive, seamless and "scalable"*:

- objects must be permanently connected to the Internet, if possible, or at least when a connection is available,

- TaaS (*Trust as a Service*) is at the heart of this infrastructure. *The Internet will be a trusted third party*;

– the digital twin of an object must combine its real and virtual aspects (be "phygital"). The aim is to have the best aspects from each space to create a much more complete and satisfying customer experience:

- a twin is constructed so that it can receive input from sensors, gathering data from a real world counterpart. This allows the twin to simulate the physical object in real time, and, in the process, to offer insights into performance and potential problems,

- a twin embodies the possibility for a program to act on behalf of a physical object to which it is attached and of which it is perfectly aware,

- a twin should evolve with each transaction,

- twins will be integrated in a *"smart wallet"*, i.e., a software application that serves as an electronic version of a physical wallet;

– *autonomy based on a decentralization* of everything:

- objects will behave like robots,

- objects are treated individually from an isolated point and are operated independently of a remote control,

- *there must be no central intelligence* controlling the totality of individual objects in a totalitarian way. On the contrary, each object is in some way autonomous and independent, with the ability to be interrogated and to interact with other network objects when necessary. In short, objects in the IoT should be *independent and communicating agents* (in the sense of AI and operating systems).

The Internet of Augmented Me (I.AM) is necessary for the business model of the IoT. The characteristics are perfectly described with the following Chinese proverb:

– *Tell me and I'll forget.* This is at the heart of *mass marketing.*

Show me and I'll remember. This is at the heart of *segmented marketing.*

– *Involve me and I'll understand.* This is at the heart of the *business model of the IoT.* Every customer is a market. The dialogue between the companies and their clients becomes real, and ethics are at the heart of the winning business models. The companies cannot lie anymore.

With the Internet of Augmented Me, the complexity of devices plays a secondary role because users work with their own digital assistants for all their daily tasks, which cover both the physical and digital worlds.

So hyper-personalization and citizen/consumer involvement are at the heart of the Internet of Augmented Me business model. The answer to the hidden expectations of the consumer citizen is at the center of value creation (contextual marketing). In this context, there will be a massive decentralization of collaboration between citizens/consumers, the AI environment (enhanced intelligence) and secure transactions with the blockchain to facilitate this involvement and training just in time.

A new wave occurs in the digital world every 10 years, with new innovations, breakthroughs, applications and new issues that concern a wider audience each time:

– 1944: the first computer;

– 1954: the first mainframe;

– 1964: the first mini-computer;

– 1974: the first microcomputer;

– 1984: the Macintosh;

– 1994: the first navigator with Mosaic;

– 2004: social networks;

– 2014: the Internet of Augmented Me;

– 2024: the brain to X (interaction of the brain with its environment).

The main features of the Internet of Augmented Me are as follows:

– goal:

- hyper-personalization and *citizen/consumer involvement* are at the heart of the Internet of Augmented Me business model;

– how:

- anticipation: answering to the latent expectations of the consumer citizen is at the center of value creation (*contextual marketing*);

– method:

- (1) key role of the *massive decentralization* of collaboration between citizens/consumers, (2) *AI environment* (enhanced intelligence) and (3) securing transactions with the *blockchain* to facilitate this involvement and training just in time;

– competitive advantage lies in collaboration and data sharing:

- *collaboration* is everywhere. It is at the intersection of humans and machines, between machines, and between humans. Humans and machines must work better together: P2P, M2M, M2P, B2M, etc.,

- *citizen/consumer access to all the services* they need, when they want, wherever they are and with the available equipment. This equipment can be a computer, a TV, a phone, a tablet, a watch, a game console or a multimedia device;

– valuable creation:

- key role of *brands*, whose valorization is the result of collaboration, which can only make sense if the parties involved are interdependent in one way or another. Their effectiveness depends on a compromise between the actors concerned to produce solutions that none of them could obtain if working independently. In this way, the actors all depend on each other to find mutually beneficial solutions. The goal is to improve the productivity of these exchanges. The PCE (*Productivity of Collaborative Exchange*) will be a very strategic indicator of the quality of this collaboration;

– business model:

- it resides in the capacity to create intangible capital and to have an apparent financial surface more important than it is, with regard to the results;

– *key role of* data and algorithms:

- the algorithms go to the data and not the other way around.

With the Internet of Augmented Me, the *asymmetry of information* has been equalized in the sense that it becomes favorable to the citizen/consumer who gains bargaining power. Tools such as blogs, forums and recommendation sites give viewers access to more information, especially the buying experience of citizens/consumers.

With the Internet of Augmented Me:

– a company sells a product, but a citizen/consumer buys a service and beyond that a UX. An object is no longer bought for its primary function but for the services it renders;

– there is a main difference with traditional business models: customers get involved and ask to be rewarded for their work;

– robot coaches and assistants of all kinds will become widespread;

– enterprises are increasingly redesigned to transform businesses in *platform companies*;

– *predictive intelligence* will be at the heart of the competitive advantages of the companies.

The success of the Internet of Augmented Me is also the result of multiple interaction opportunities between companies and their customers, which will create *new challenges for companies*:

– companies must synchronize their communication across all channels and deliver the most relevant content for each of their citizens/consumers at the best time and through the most effective channel;

– they must reinvent the customer relationship and put the consumer at the center of their strategy;

– they must transform unique transactional relationships into a long-term, permanent interaction with more frequent interactions;

– they must get a recurring revenue stream, while customers exchange a single capital expense for a tailor-made solution that can be increased or reduced as needed, while gaining support and quality service;

– they must reason in terms of flows, exchanges, interactions, collaboration and not stocks, which means, in particular, that a company does not own its employees, suppliers and customers.

The IoT facilitates this transition by adding sensors to its products as well as accurate, cost-monitoring devices. This provision of historical and real-time data enables the creation not only of automated alerts and preventive maintenance, but also of predictive maintenance and predictive well-being.

We will move from the traditional product-based approach to a marketplace approach. The goal is to deliver optimal value to consumers via multiple ecosystems that will interact with each other. The goal is to interact with the consumers so that the producers will understand hidden expectations and create value after the sale of the product.

In this context, the value of an object is getting closer to 1 euro. The creation of value is the result of the implication of the citizens/consumers.

In this context, the subscription-based business model makes obsolete the loyalty programs of retailers based solely on prices. It is a way to enter the daily lives of consumers. Subscription can therefore be the gateway to services that encourage the purchase and use of new products. A good example is that of Nestlé, which offers a Nespresso machine against a subscription that will soon be a piece of the puzzle, alongside other channels such as personal assistants.

In data wars, subscription is a way to collect valuable information so that companies can better target customer expectations. Faced with the power of GAFA, it offers a competitive advantage to brands and retailers.

This new way of making business implies that more and more data must be collected from each transaction or process in real time. Any device equipped with processors can be connected to a network and can start broadcasting large amounts of information.

The major challenge for companies is therefore to understand how these interactions with citizens/consumers will create a competitive advantage for pioneers by focusing on network effects, and therefore on the Internet of Augmented Me.

1.2. Characteristics of the Internet of Augmented Me

It is important to understand the growing significance of seven concepts that are inextricably linked to the Internet of Augmented Me, in order to fully comprehend I.AM. technologies:

1) new way of building trust, Trust as a Service;

2) (digital) twin of everything (TWE);

3) interaction with objects, ATAWAD (AnyTime, AnyWhere, Any Device);

4) massive decentralization of everything and autonomy of objects;

5) unification of all the media (Unimedia) so that users will only have one interface and service continuity;

6) anonymous personalization;

7) robot coaches and predictive/emotional intelligence.

1.2.1. *Building trust and autonomy*

With the Internet of Augmented Me, an organization can be represented as a set of interactions that flow between internal actors and between the organization and its external stakeholders.

An organization can only live sustainably if it is perceived as creating sufficient value not only by the capital but also by current and potential customers, staff, suppliers and the society in which it operates. Otherwise, the conditions for sustainable viability are not met.

In this perspective, there is more value than that perceived by the "client"; it is at the heart of the Internet of Augmented Me. The true value of an object is not its cost price but the price of the service and user interface that the customer is willing to pay for. For the Internet of Augmented Me, the value of the service is not the value that the creators of this service decide upon, but the value that decides "Me".

Remember that the customer does not buy a product, but buys the services associated with it. Moreover, the services result from relevant interactions between objects, ideas, people, businesses, organizations, etc. These interactions must be based on:

1) *emotion* (because every decision has an affective dimension);

2) *communication* (I only buy a product or service if I know it exists);

3) *understanding* (everything must be clear and transparent);

4) *trust* (I buy in the hope of being satisfied).

The four words: emotion, communication, understanding and trust summarize the major trends of the new digital innovations (anthropomorhic view).

We must therefore abandon the logic of producer or specialized distributor and enter into the logic of the customer, in order to understand and anticipate their contextual expectations: the offer must become global and provide consistent solutions to all questions that a customer may raise in a given situation.

The consequence is clear: with the Internet of Augmented Me, we go towards a predictive hyper-personalization.

The key to the success of this business model is the *gamification* of any business (BaaG – *Business as a Game*) and the marketing of the good news. Gamification will be one of the major themes of the coming years. The difference is that today it feeds on advances in artificial intelligence and augmented reality.

The next step is *real-time marketing*, which is the reward of citizen/consumer involvement. This type of marketing will allow customers to be targeted even more finely and improve ROI campaigns.

By designing a bike with a screen connected to the Internet, broadcasting live cycling courses in their studio, Peloton (*https://www.onepeloton.com*) revolutionizes fitness cycling. This bike is equipped with a 19-inch screen connected to the Internet, which allows riders to follow live courses led by senior instructors in their studio. This unique feature allows runners to participate in gamed classes where they can track their progress on the class leader board and can sometimes even be called by live studio instructors.

Unlike other bike manufacturers, Peloton does not just sell equipment. Using the commercial "razor and blade" model popularized by Gillette, Peloton sells "razors" (stationary bikes) and "blades" in the form of monthly subscriptions to online courses, in addition to an increasing number of new sources of revenue for a captive audience. The monthly subscription model gives users the right to participate in online courses as well as other courses, ranging from strength training to the high-intensity stretching and interval routines.

Peloton also completes its business model by offering clothing and fitness equipment.

Under Armor (https://www.underarmour.com), in its record application, uses AI to collect health information about physical activity, sleep and nutrition to make personalized recommendations about workouts and health goals. The app also serves as a personal trainer and can analyze the workouts to make sure they are as effective as possible. Under Armor's use of AI helps clients get the personalized health recommendations they are looking for.

Box 1.1. *Example of Peloton and Under Armor*

We are moving from 360-degree communication, where the purpose was to occupy the spaces of communication, to a real-time conversation with the consumer, to get the right message and the right offer, at the right time and through the right channel.

If real-time marketing offers multiple possibilities of interaction with customers today, companies face new challenges:

– they must equip themselves with technological tools that will make it possible to analyze the data coming from more and more channels;

– they must synchronize their communication across all channels and deliver the most relevant content to each of their citizens/consumers, at the best time and through the most effective channel;

– they must reinvent the customer relationship and put the consumer at the center of their strategy;

– they must protect and value their customer data.

In this context, companies need to build their strategies around data and trustnology[1] (technological innovations being at the heart of building trust), and therefore transactions.

It is only by industrializing the implementation of these innovations that companies will be able to define and manage their business strategy, making full use of the data they have at their disposal, thus developing better self-awareness, as well as their environment. In this context, the marketing of intention and emotion will become increasingly important.

Each company function will be connected to the customer – from the design to the product marketing strategy – so that companies will have to adapt quickly pto integrate the connected objects and customer data generated by the IoT, in order to optimize relations and citizen/consumer involvement. *The aim is to create an environment of trust* between citizens/consumers and businesses.

To create this environment of trust, it is obviously everyday tools, first and foremost, that will be connected to the Internet, such as smartphones, tablets, laptops or televisions. However, it is the objects of our daily life, in particular, that will create new distribution channels: surveillance cameras, cars, glasses, light bulbs, household appliances, thermostats, scales, etc.

1 Uniting trust with technology.

1.2.2. *The digital twin*

Wikipedia defines the digital twin as follows: whether through a mobile health application or a connected scale, the uses of digital twins are based on data capture which is increasingly automated and induces the circulation of large masses of personal, sometimes intimate, data. These data exchanges are at the initiative of the individuals themselves, provided they wish to share their data.

In other words, the objective is to keep a precise (quantified) account on particular elements of one's life, in order to keep track of it over time and improve one's performance. This is based on principles and methods that allow everyone to measure his/her personal data, analyze it and share it.

The digital twin will measure and quantify everything that happens about "oneself", in terms of health, and also in the context of one's professional activity (number of emails sent) or personal activity (number of friends added on Facebook), in order to highlight their most significant elements. Almost all projects of quantified self are summarized today to visualize the data.

With a subscription, you can have access to live and on-demand classes such as yoga, boxing and cardio or bar sessions, all led by certified instructors.

The principle is simple: we simultaneously look at these instructors who move on the mirror and evaluate our own action based on what is done by the instructor on the mirror. The instructor is in a remote studio.

Box 1.2. *Mirror example*

It is also possible to measure many other aspects of people's lives, such as sleep patterns and productivity.

A digital twin is also a computerized companion of a physical object that partially models it (from a simple name to a sophisticated 3D model) and uses sensor data installed on it to represent its state and position almost in real time. Someone who is not in the physical world of the object could still (remotely) follow its evolution by exploring the digital world.

According to Gartner, in the next three to five years, hundreds of millions of objects will be represented by digital twins. Organizations will use digital twins to repair proactively, plan manufacturing processes, run factories, predict hardware failures, increase operational efficiency and improve product development.

An example that has existed for several years is that of Babolat's connected racket: Babolat Play. The internal sensor located in the racket's handle measures the vibrations of the racket during a match and represents, in real time, all ball impacts, angles, stroke directions (forehand or backhand). The company's challenges are to improve the performance of objects, equipment, systems or factories.

A good example is that of small equipment that is worn on a person which informs them about the evolution of their performance. Such devices are already commonly used in sports, such as cycling or jogging. They inform the cyclist or the jogger about their heartbeat and assist them as a coach would do.

Thousands of connected objects will be represented by digital twins. Physical data can be used to provide information on how the components of an object work and respond to the environment, as well as data provided by sensors in the physical world.

1.2.3. *The interaction with objects, ATAWAD (AnyTime, AnyWhere, Any Device)*

Always being connected is not a trend, but a reality firmly anchored in everyday life that has influenced many buyers for several years.

The ATAWAD acronym means that individuals can connect at Any Time, AnyWhere, with Any Device. Using it acknowledges the increasing mobility of people and the quasi-ubiquity made possible by information technologies. The term "mobiquity" (ubiquity in place of mobility) combines the two notions.

More concretely, ATAWAD refers to the ubiquitous marketing or mobiquity, or the real-time, precision marketing. With ATAWAD, the consumers/citizens can access a service when they want, where they want, with the devices they want.

1.2.4. *Massive decentralization of everything*

The progress in communications and computing power ensures that all objects become capable of capturing and processing data, and of communicating with each other and with us.

The world can be described as a set of interconnected networks at all levels. These networks have different abilities, some are limited to transmitting messages and orders, such as our nervous system, and others convey data flows with very high throughputs.

This massive decentralization of everything requires measuring the PCE (Productivity of Collaborative Exchanges). This indicator will be at the heart of the new way of doing business.

In this context, a logic of communication exists between peers and between equals, as opposed to that of a pyramidal, centralized organization, as in the majority of companies and administrations. Another important factor is the one that facilitates communications, sometimes to the point of causing a confusion of sorts.

In a world managed by blockchains, decentralization could be pushed even further to include objects. Once they have their own identity, they can be controlled via a blockchain and will, in some way, be self-determining.

1.2.5. *Unimedia and open marketing*

The unification of all the media on the same platform means that all modes of expression and all products of the human mind can be translated, stored, transformed and transmitted in a digital language. The boundaries between markets and trades are blurring while creating new competition.

The example of the scale is a very good illustration of the Unimedia concept. Now, the user of a scale does not buy an object to measure their weight, but a service enabling them to reach their goal: to gain or lose weight. The user therefore expects "unified" service proposals of the type:

menus, sports, dietary advice, etc. They will understand very well that the scale is connected, for example, to the refrigerator, thus becoming capable of providing a concrete answer to the following demand: *Here's what I want to do, do whatever it takes.*

Another example is that of the car. Today, a car is more than a way to get from point A to point B. It is a center of activity within everyday life. The user tries to have truly connected experiences in their car, to do more, to save time, and to make life easier. The overall goal is to make a person's workday more efficient.

The width of the touch screen of the Byton electric car is equivalent to the width covered by seven iPads when they are arranged side by side, or the same as 24 smartphone screens.

Called "Experience Shared Display", the touch screen can be controlled directly or via a 7.0-inch touch screen placed at the center of the wheel or even via an 8-inch touch screen in portrait placed along the center console.

Each of these screens has its own interface for entertainment, phone connectivity, navigation, air conditioning and Cloud services. It is also possible to hide these screens and use them as remote touchpad controllers on a 48.0-inch screen.

The main screen not only displays speed, navigation and other *infotainment* functions, but can also be used by passengers for chat rooms, video games or health information.

The system also has the ability to recognize gestures made near the center of the dashboard and to respond to voice commands.

The business model of Byton will consist in not only selling cars but also in using the car as a platform. The goal of Byton is to earn more money, on the one hand, through the sale of digital content and shared mobility, and, on the other hand, from its own activities such as carpooling and car sharing (rather than just linking its vehicles to existing services such as Uber, Lyft or Waymo).

Box 1.3. *Byton example*

The cars of tomorrow will be the result of unified heterogeneous resources from different vendors operating in the logic of ATAWAD and Unimedia:

– unified management data from the car and its environment;

– centralized monitoring to improve the efficiency of operation and maintenance;

– response to the latent expectations of car users in a logic of continuity of services.

It is neither an embedded operating system nor a "finished product". The Unimedia concept implies a lively and agile platform based on the Cloud and aims to address the five main scenarios: (1) predictive maintenance, (2) improvement of the car in productivity, (3) advanced navigation, (4) customer knowledge, and (5) aid to the creation of autonomous driving capabilities. Unimedia connects citizens/consumers seamlessly no matter where they are located.

The example of the building is another excellent illustration. A Unimedia interface will enable unified interconnection and interworking between various control devices in smart buildings, management of tens of millions of IoT terminals, as well as distributed deployment and flexible extension of intelligent building management. This interface will simplify interoperability and minimize the integration costs of an intelligent building.

The goal is to provide a variety of IoT interfaces that can be used in a wide variety of intelligent building access scenarios, or for unified access and management solutions to subsystems, such as lighting, security, power distribution and networks, metering, heating, ventilation and air conditioning (HVAC), and electrical control.

In this context, the robot coach is the indispensable complement of the Unimedia.

1.2.6. *The robot coach*

The "robot coach" will be more and more likely to make decisions that could have been made by human managers. Concerning supervision functions related to monitoring employee performance, which is measured by quantitative data, better and faster decisions will be made by intelligent machines.

Artificial intelligence (AI) and machine learning (ML) will be at the heart of this assistance with, in particular, foresight. The use of AI will help improve user experience and streamline business processes.

The goal is to move towards robot coaches and learn as much as possible about people's lives, to get all the relevant data on their daily lives, and then automatically suggest relevant apps, based on location and time of day.

Stuffed with artificial intelligence (AI), robot coaches will try to stick, as much as possible, to the natural functions of the millennial human in order to appeal to our senses, to mimic the behavior of the brain, and ultimately to complement and increase it.

Voice, gesture in space, body movement, gaze (like a laser pointer): with the new vocal and gestural interfaces in 3D, computers, digital devices and the Internet are getting closer to our bodies, so that our bodies will be augmented.

More intuitive and, above all, much more personal, this computer science is also close to psychology, cognitive science, feelings, discernment, reasoning, decisions, our perception and therefore our relationship with the world.

We are now talking to machines that respond to us. Artificial intelligence will be used on demand, and virtual worlds will become more and more realistic, due to the increased power of computers, graphics cards, optical systems and tomorrow, glasses.

In summary, the new era of "post-smartphone" and "post Google" perceives, understands, learns, reasons and predicts. It will transform our organizations and our daily lives, change the way we interact with our terminals or the Cloud, and promote a real human/machine co-production, leaving to the machine the thankless tasks. It will interpret natural language and learn from previous uses.

By using knowledge-based versions that depend on neural networks, the Internet of Augmented Me will prevail everywhere. IoT infrastructures will play a key role in assisting citizens/consumers and responding to events. In this context, the systems Blockchain + AI + IoT will "act" in the place of citizens/consumers, proactively.

Macy's (www.macys.com) uses artificial intelligence to personalize the store experience. When a customer enters the store, they can start chatting with a digital assistant on their phone. The chatbot asks questions to guide the customer's shopping experience, and gives recommendations and directions for store items. If the bot feels that the client has become frustrated, it will warn a human partner to regain control as soon as possible.

Finding the perfect jeans can be a frustrating experience, but Levi's (www.levi.com) recently launched an AI-compatible chatbot to help customers find the perfect pair. Virtual Stylist uses natural language to determine what each customer is looking for in a pair of jeans. The bot helps to discover the lifestyle and style preferences of each client. This information is combined with the training that human stylists receive in order to provide accurate recommendations. Once the chatbot has recommended jeans, customers can share this information with friends to get their opinion.

Box 1.4. *Example of Macy's and Levi's*

In this context, instant messengers and their automated bots are becoming the main entry point to the Internet, to the point where the relevance of a site or even an app now arises.

In the long run, couriers, meta-platforms and new browsers will become the only points of entry. They have already supplanted social networks in our uses, serving as a basis for these new conversational services, which, through grammatical analysis, semantics and speech recognition, recognize what you say and what you think.

Recall that bots are software programs that automate everyday tasks using more dynamic discussions, so that traditional static apps need not be installed. The bot does what it is asked to do: *Here's what I want to do, do whatever it takes!*

Bots have search engines and will therefore give quite a blow to old apps by using them to combine the advantages of the web (immediate access without having to install anything, link management) and native applications (user identification, access to user context), without their disadvantages (notably, the need to acclimatize/adapt each site or app).

Mobile equipment will become real robot coaches or personal assistants that are able to hold discussions to perform more and more tasks: inform us, book a plane, a hotel room, order a meal, a taxi, pay an invoice, play, advise on an investment or organize the agenda. Without even typing a text, searching, clicking, opening an app, switching to another, finding its password, etc.

More generally, according to the Ericsson study, artificial intelligence will replace smartphones within five years.

Artificial intelligence (AI) interfaces will take over, replacing smartphones in five years because the interface will be vocal so that the user will "chat" with the machines.

Robot coach is a good example of what will become mainstream. A robot coach named Roy now takes care of your time management. With AI, he is able to schedule appointments and tasks in the blink of an eye. In doing so, he considers many other factors, such as whether the appointment is work-related, whether you have enough free time, what time of day is best for you, whether the weather is nice, and much more.

A robot coach can not only plan cleverly for you, but also show you the best restaurants in the city, call you an Uber for your next appointment, find time for your weekly workouts and learn from your preferences and habits. From now on, you have your own personal robot coach.

1.2.7. *The unification of communities*

Communities are fragmented networks. The Internet of Augmented Me provides an opportunity to unify these communities based on the connection and collaboration of its participants with its ability to connect, coordinate, process and categorize, on a large scale, the information needed for every citizen. The problems are so complex and so vast that they cannot be solved by any centralized organization. They must be solved in a decentralized manner resulting from the increased productivity of collaborative exchange within each community.

This unification of communities is based on relationships, attitudes and a culture of collaboration whose objective is to value the common good. The business performance of companies depends on the quality and the effectiveness of their services and how they interact with their customers in those communities.

1.2.8. *Anonymous personalization*

With the Internet of Augmented Me, predictive hyper-personalized services will be generalized with a guarantee of anonymous customization (like a Swiss bank) to follow all of our activities in a globally digital and massively interconnected world: measuring our sleep cycles, counting our steps, our sports hours, listing what we eat, and processing this information.

1.3. Towards business platforms and subscriptions

The essential role of the platforms is obviously to manage the data coming from billions of connected devices that will soon provide a continuous flow of real-time data, used to reinvent traditional products. These platforms will become *service supports*. This requires a physical infrastructure to collect the data that will be needed for technical intermediation.

With platforms, it is important for companies to design *predictive applications* that can sense their environment, respond in real time, anticipate the actions of the user and citizens/consumers, and meet users as they need it in the logic of products/services described above.

The big changes in the new way of doing business will come from the production, the distribution of these objects, and appear especially in the relationship with citizens/consumers after their purchases. Production is a separate process that ends when the product is shipped to the consumer. Connected products cannot operate without their Cloud or SW-enabled edges, without an operator, and therefore without the producer/distributor of the product and without the services associated with connected objects.

In fact, technology stacking is a component of the product, which the manufacturer must continue to manage and improve throughout the life of the object. In this sense, production becomes an ongoing process, especially maintenance. The price of the object must fit into this service logic. In other words, the price of the object is 1 euro, whatever the object.

More generally, objects will be increasingly perceived as the materialization of the service (the physical part of the phygital). The integrated platforms, hardware and software bases on which these connected objects operate will be the essential complement to "connected" objects.

They will radically change traditional industries that are deeply rooted in economic models and usage patterns that have not changed for a very long time.

> The feet of children are constantly growing. The need to buy them expensive shoes is a thing that lasts forever. In an effort to simplify the shoe shopping experience for parents, and to reduce costs, Nike gives them the option of ordering shoes for children aged 2–10 years.
>
> Twice a year, Nike will send a prepaid bag in which parents can return their children's worn-out shoes, which will either be donated to families in need if they are in good condition or recycled via Nike Grind, a program that separates the rubber, foam, the mixture of leather and textile, grinds them into granules and incorporates them into new products, including shoes, clothes and playing surfaces.

Box 1.5. *Example of Nike*

As before, advances in the IoT allow machines to communicate not only with humans (citizens/consumers, collaborators, etc.), but also to work side-by-side with humans. The objects are only the materialization of the service.

Example of the scale: we do not buy the scales. We buy the service to lose weight and improve our wellness via dietary advice, special food, sports, activities, etc.

This proliferation of connected objects will require a cultural change, such as the one that includes the maintenance of real-world objects and collaborates with data and IT professionals.

With the Internet of Augmented Me, companies must build their strategies around data.

1.4. Examples of business sectors that will be impacted by I.AM

Commerce is an excellent example of the profound changes occurring in the way of doing business in the era of the Internet of Augmented Me, with the key role of valuing the common good.

For example, OpenStreetMap is an open alternative to Google Maps, with an entirely different business model. The access is guaranteed and free, with a free reuse in respect of the license. Durability is insured, since one can copy, preserve and distribute all the data. This forces OpenStreetMap to find other sources of funding and other forms of partnership, and to manage a large community of contributors (individuals, businesses and communities).

The trade will be more effective as consumers will be able to express their desires in full trust, individually, anywhere and anytime. This new situation is shaking up social hierarchies because it provides a decisive advantage to experienced (agile) users and significantly reduces the importance of knowledge in personal decision-making.

Meanwhile, the experiential side of the retail business will change, while the part of discovering good products and socializing with others will not go away. Citizens/consumers will continue to visit physical stores, not because they want to "buy things", but because they want experiences which they will not find anywhere else.

The impact of technological innovations will be measured by the ability of merchants to predict our behavior and preferences, as well as to provide *automated delivery of everything* that is consumable, be it bread, toilet paper or a replacement phone charger, and this *just in time*.

The retail trade will be increasingly dynamic and diversified. There will be more independent shops, not fewer. There will be more craft products. There will be more products to be more carefully selected. It will be notably easier to manage a retail store.

When buyers buy tangible products (beyond the grocery store, for example), they will want them to tell a story, to fit into their lifestyle, to be completely unique, or all of the above.

Allowing customers to create and customize products not only meets these standards but will also make shopping much more exciting. The technology will allow people to build their own purse or their own watch.

Companies such as Amazon are investing heavily in robots to help pack and ship items. Amazon already uses robots in about 20% of its execution centers. We will also see more robots online, in the form of chatbots.

As more and more consumers embrace instant messaging applications such as Facebook Messenger or WhatsApp, we can expect retailers to use these platforms to talk to customers and streamline communications.

For example, if a customer wants to track their order, they can simply “ask” the reseller on Messenger, and then a chatbot can automatically retrieve the shipping information. Other retailers (e.g. Victoria’s Secret) use chatbots to showcase their product lines.

Another major application of these innovations is to allow merchants to have as much information about their customers and their spending and consumption habits, and to build a relationship of trust with them. In other words, it is important to go beyond the transactional aspect to the personal aspect, and to make data and analytics a competitive advantage for retailers and brands by together building an ecosystem of shared intelligence.

Another important point is to be able to overlay digital business data with stored physical data (phygital business), allowing for deeper analysis of buyers and customers.

For this, *augmented reality* (AR) will play a key role. It has been around for a while, but it will become more sophisticated.

With the implementation of the digital innovations described below, physical stores have a bright future. Traditional store formats may be declining, but the innovative stores – those that offer shopping experiences that focus on e-proactive trade – will have huge success.

Technology will accelerate physical store opening. With point-of-sale management systems, it has never been easier to open a retail business. For example, mobile applications and Cloud-based solutions will allow merchants to quickly install with minimum investment. For this reason, we will see more and more independent stores entering the market.

With digital innovations, advertising costs have never been lower because of the power of social media. Independent stores may develop a relatively low-cost customer base if they offer a product or service that attracts a targeted customer base. Of course, access to software and hardware to empower sales will also be cheaper allowing small retailers to develop a connection and understanding of what their customers want.

As technologies such as Big Data and machine learning (ML) continue to evolve, retailers will need to continue to look for ways to collect and leverage data in their sales, marketing, customer service and their operations.

Forbes recently reported that Walmart is already using ML to optimize delivery routes for their associated home deliveries. Walmart uses facial recognition, for example, to tell which buyers are dissatisfied and which are not. This data can then be used to tell associates to open new paylines, which is just the beginning.

Successful traders are those who can successfully evolve by involving citizens/consumers. In any case, it will put the customer in the foreground and create a more personal store experience.

The car is also a good example. As autonomous cars evolve, the need to drive the car or the fact to focus on driving disappears or becomes minimal. As a result, the need to have a car designed around the act of driving will likely be overshadowed by the need to design the "car" space and how it will be used. The car of tomorrow will make the city of tomorrow.

The house will also be completely transformed by the Internet of Augmented Me. Indeed, as appliances such as our TVs, lamps, thermostats, garages, cooking appliances, coffee machines and robot vacuum cleaners will connect to the Internet, *we will need home management services*. These devices will be smart enough to talk to each other, but we will have both a management layer and an interaction layer at home (or in the Cloud).

As the technology such as HomeKit, Echo and individual devices become smarter, such resource management goals and objectives will have some major features, for example to:

– effectively manage the house;

– customize the home environment to suit your tastes, preferences, etc.;

– respond to events and requests in real time if needed;

– learn;

– inform.

The *intelligent city* (*smart city* or *augmented city*) is therefore first and foremost a long-term planning mission, for architects and urban planners, and especially for local authorities with a clear objective: technologies must meet latent expectations of citizens who should no longer be seen as consumers of services but as partners and stakeholders in the economic development of the city.

The city will also be organized around the Internet of Augmented Me. Indeed, the success of a *smart city* depends primarily on increased interaction with the citizens/consumers.

Obviously, the industry, which will naturally be impacted by the Internet of Augmented Me, is health/well-being. Recent technological developments in mobile health monitoring systems (mHealth) are increasingly incorporating portable sensors and algorithms that will track a patient's vital signs. In this context, personalized medicine will emerge, and smart drugs will no doubt be part of the treatment's hyper-custom that will radically change the behavior of citizens so that health problems are anticipated.

1.5. Business model of the Internet of Augmented Me

1.5.1. *Reminder of the definition of a business model*

The *business model* is a tool that describes how to create, deliver and capture value for a business. It must allow companies and managers to understand the operation of a company.

The model shows the dynamics of some companies that have proved more viable than the large monolithic vertical companies of the past. Over time, a business "splits" into three fundamentally different business areas. These three entities can coexist in the same organization, but it is better if they are different entities to avoid conflicts and unwanted compromises.

The three entities naturally play different roles:

– *customer relationship* identifies, attracts and builds relationships with customers;

– *product innovation* designs new products and services and monetizes them;

– *infrastructure management* focuses on building and managing facilities for high-volume repetitive operational tasks.

The aim is to identify their resources and capabilities and to adapt them to the needs of the market. It also serves as a model for carrying out the business strategy.

1.5.2. *Main features of the business model of I.AM*

The characteristics of the Internet of Augmented Me are very different from traditional models that are essentially based on anything that can help push the product. With the Internet of Augmented Me, it is the opposite. The business model is based on the response to the latent expectations of consumer citizens and forecasting intelligence.

The PER of an "Internet of Augmented Me" company is not that of an industrial company watch.

We go from the profitability of a company based on MATERIAL capital to a company based on IMMATERIAL capital.

The profitability of the Internet of Augmented Me enterprises is not that seen by the rational actors. It is measured by the ability to extract value from the QUALITY of interactions with citizens/consumers to create CONFIDENCE and therefore the ability to create a TRUST capital.

This new form of profitability will generate the valuation of the company, which can reach considerable sums.

The business model of the Internet of Augmented Me is therefore based on its valuation and not on its results.

With traditional economic models, we are talking about added value and the 80/20 law.

Several conditions are required for the actors of the ecosystem of the Internet of Augmented Me, some of which are as follows:

– It is important to reason in terms of the effective use of digital for ALL devices that interact with citizens/consumers at all times of everyday life in a globally digital and massively interconnected world, not just in terms of equipment. These tools have many features and competition is now organized in an ecosystem according to how they know (more or less) to implement the features that create value for citizens/consumers.

– More concretely, digital innovations no longer only transform the world of production (the factory) or the world of management (offices). They transform the market and non-market exchanges with citizens/consumers. It should be noted that the economic sectors where the differences between the

United States and Europe are concentrated are sectors such as retail trade, wholesale trade, the sale of airline tickets and the brokerage of savings products and insurance.

– Objective: to ensure continuity of service and orchestration of these services in all moments of daily life in a globally digital and massively interconnected world.

– Technological culture and performance measurement issues do not arise in the same ways as in traditional enterprises. It is not just about culture. It is a matter of user experience over a long period and therefore of understanding the latent expectations and needs of citizens/consumers. Companies no longer need to define their business by what they sell, but by what the citizen/consumer wants and therefore buys.

– There is also a question of substance: what can be measured out of relevance in all those activities that are at the heart of everyday life in a globally digital and massively interconnected world? If we do not measure, we will not be able to control the relationship with the citizens/consumers and with the partners of the company.

– It is clear that it is not enough to control digital investments by their cost and that we cannot just measure the productivity of work. We must control the uses to reduce the cost of possession and to increase the use value. Predictive intelligence is at the heart of effective steering.

– In short, we must transform the business relationships with suppliers and customers, digital technologies that are more deeply involved in the transformation of the value chain than would a simple productivity tool, but a tool that aims primarily at competitiveness.

A central indicator of the Internet of Augmented Me is the measure of ROCE (Return On Capital Employed). It is measured by the ratio between the operating profit of the enterprise and the total capital committed to produce this result: fixed capital plus current capital. What we must measure is first and foremost the impact of digital investments on the company's performance.

We will move to a "zero everything" business model. The goal is to sell before producing with zero stock, zero energy and zero waste. How? By focusing on the Internet of Augmented Me or on the involvement of citizens/consumers and the PCE (Productivity of Collaborative Exchanges).

This indicator is all the more important since, when compared with the weighted average cost of capital, it allows a company to know whether it is in a situation of creation or destruction of value. Today with the Internet, digital technology not only affects the productivity of work, but even more so the profitability of capital: office occupancy, rotation rate of trucks, trains, cars, planes and stocks.

Several companies believe that digital transformation saves time and money, gaining a competitive edge and appealing to customers with sophisticated applications and innovative technologies.

They are wrong. Digital transformation is about improving the customer experience. And many companies are finally beginning to understand that customer experience management needs to be taken into account if they want to prosper over the next few years.

The elements that structure the business model of the Internet of Augmented Me can be summarized with an adaptation of Maslow's pyramid, as shown in Figure 1.1.

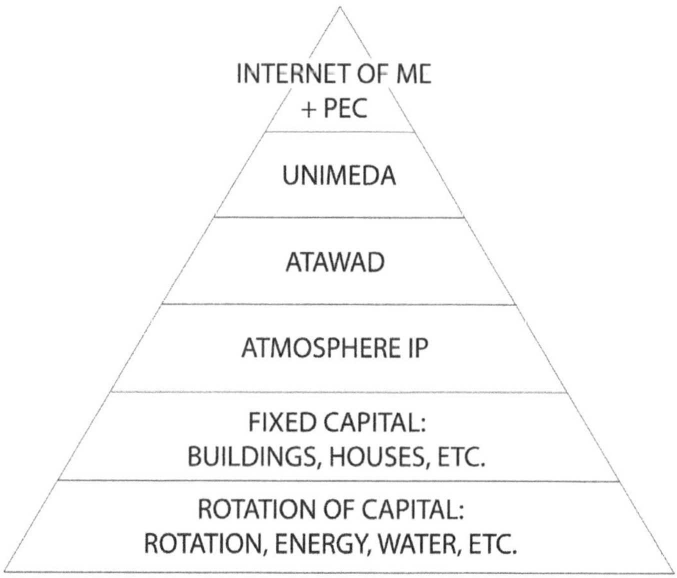

Figure 1.1. *Adaptation of Maslow's pyramid*

The model therefore shows the dynamics of some companies that have proved more viable than the large monolithic vertical companies of the past.

To summarize the points thus far, a company "is divided" into three fundamentally different areas:

– customer relationship identifies, attracts and builds relationships with customers;

– product innovation designs new products and services and monetizes them;

– infrastructure management focuses on building and managing facilities for high-volume repetitive operational tasks.

The method proposed by Osterwalder and Pigneur[2] includes nine building blocks that illustrate the logic of the way in which a company intends to make money.

The objective is to formulate and structure the business model and thus obtain a holistic vision of the main elements that structure the business model of the Internet of Augmented Me.

The nine building blocks of a business model are as follows:

– segmentation of the company's customers;

– relationships with the customers of the company;

– distribution channels offering products and services;

– value proposition;

– main partners of the company;

– customer adherence to the values of the company and its brand;

– key resources of the company;

– cost structure of the company;

– company income.

2 https://www.strategyzer.com/books/business-model-generation.

1.5.3. *Customer segmentation*

– Segmentation can be based on common needs, common behaviors or other attributes. Clear segmentation is essential for the company to create value.

– With the Internet of Augmented Me, every customer is a market.

– The segmentation of the company's customers therefore has only one purpose: to optimize the customer experience and thus the creation of an appropriate internal environment for employee engagement. We will talk more about segmentation of experiences than about customer segmentation.

– Customer segmentation should help proactively focus on a high-quality customer experience that can affect the entire lifecycle of the products and services of the business.

1.5.4. *Relationship with company customers*

– The modes of interaction between the company and the customers deeply influence the user experience.

– In the digital age, the consumer is no longer sensitive to mass advertising. They are more receptive to personalized advertising and therefore to products that meet specific needs.

– With the Internet of Augmented Me, citizen/consumer involvement is at the heart of the business model. The quality of interaction with citizens/consumers is a key element in building trust. The good question that every business should ask is: How many people have noticed my offer and asked to interact? The whole company is focused on the "customer" experience and its co-creation.

1.5.5. *Distribution channels of supply, products and services*

– It serves as an interface between the company and the customer, who plays a vital role in the user experience.

– With the Internet of Augmented Me, short circuits are needed. We must eliminate all the intermediaries that do not have enough added value. These distribution channels include SEO, social media and especially content marketing. For example, content can be used as a way to contact key players in your industry that you would like to have as business partners.

1.5.6. *Customer adherence to company values and brand*

– The consumer/customer will become an influencer because they have the experience and can testify to it.

– With the Internet of Augmented Me, businesses will have to take into account the work done by citizens/consumers to increase the attractiveness of the brand. The goal is to reward the customers for their work.

– The remuneration of this work could be done through cryptocurrencies, for example, which would be the sign of adhesion to the community.

– This notion of community will become very important because it is fundamental.

> Harley-Davidson customers can be considered the most loyal in the world. They are the best ambassadors/influencers of the brand. Some even go as far as asking for the company's logo to be tattooed on their bodies, which displays steadfast dedication to the motorcycle manufacturer.
>
> For this type of loyalty, Harley-Davidson creates a community for its loyal customers, organizing social and educational events where customers with a common interest can meet, discuss and learn about the product and share stories. Harley-Davidson does not have loyalty points. Their strategy is built around the best possible customer experience.

Box 1.6. *Example of customer adherence to a company brand*

1.5.7. *Value proposal*

– The value proposal is the glue that unites all the blocks of the business model.

– The value proposal therefore describes the fundamental value of the product or service that can be passed on to customers. The value proposal usually shows in which dimension a product or service can stand out among its competitors.

– The value can be quantitative (e.g. price) or qualitative (e.g. the value of the brand). It describes the most important tasks that a company must perform to make its business model work. These strands cover a wide range of areas such as software development, supply chain management or consulting, according to different business models.

– With the Internet of Augmented Me, the main value proposal is an anticipation or a response to the latent expectations of citizens/consumers. The company does not sell a product or service. The company sells what its customers buy from it over a long period.

– Ethics is a key element of the business model. We do not lie. The goal is the potential for large data collection.

– The value proposal is based on what the customer is buying in context and not on what the business is selling.

– With the Internet of Augmented Me, personalization is essential. Customer information should help companies make better decisions about their efforts in personalized relationships with customers and provide the necessary information to help their "distributors" to sell more and better. The goal is simplicity and therefore unification and service continuity.

1.5.8. *Main company partners*

– The objective is to describe the network of suppliers and partners required by the trade. For example, some activities are outsourced and some resources are acquired outside the company.

– With the Internet of Augmented Me, advertising is being replaced by expert influencers who are a key part of companies' competitive advantage. The partner ecosystem needs to network in a decentralized way.

1.5.9. *Key company resources*

– Key company resources are the most important assets required by the business model. They can take different forms: physical, intellectual, human or financial, depending on the different business models.

– With the Internet of Augmented Me, the company must rely on digital innovations to gain a competitive advantage. It must have the best expertise in the IoT, artificial intelligence, Big Data and blockchain.

1.5.10. *Business cost structure*

– The business cost structure describes the most important costs associated with operating a business model. Some business models are cost-based and focus on minimizing and optimizing overall costs.

– With the Internet of Augmented Me, the massive decentralization of the company is essential to make the company flexible, adaptable to its competitive environment, and to let partners bear their costs. BTO (Build To Order) without stock is also a key factor in the company's competitive advantage with resilience.

1.5.11. *Company income*

– The revenue model must of course be aligned with the interests of users.

– It describes the main revenue streams that the company receives from the business model. They can come from one-time transactions or from predictable future incomes (as subscriptions). The difference between cost and income in each case determines the overall profit of the business.

– In the business world, growth is often confused with profitability. This is not the case. With the Internet of Augmented Me, companies will achieve very high growth rates and therefore increase their value. This does not imply that they are profitable, especially in the short-term.

– With the Internet of Augmented Me, the remuneration of a captive audience is an important differentiator, unlike with a Google-type model that monetizes targeted audience or with that of Facebook which monetizes "direct" marketing.

Emergence of collaborative business models and services

Value creation for customers will move from equipment to services that will increasingly be based on the previous or expected behavior of those consumers. The actual operation of the algorithms should not be a black box. This operation will have to be explanatory. It will reduce the TCO (Total Cost of Ownership) and increase the TVU (Total of Value Usership). The goal is to move equipment into infrastructure whose apparent cost is almost free. The added value will be more focused on services over a period of time and therefore in the subscription of consumers.

Emergence of the Empowerment of Me

Consumer involvement will play an increasingly important role. This involvement will have to be rewarded/paid for as work that contributes to increasing the value of a common good.

The goal is to create an environment of trust that reduces transaction costs.

The levers and enablers of I.AM

The levers depend on the ability of companies to gain a competitive advantage by leveraging collaboration and creating value after selling equipment with short circuits. It will be at the intersection of humans and machines, between machines and between humans. Humans and machines need to work better together to generalize BTO (Build To Order). The PCE (Productivity of Collaborative Exchanges) will play a very important role and its implementation will depend on the mastery of digital innovations such as Big Data, artificial intelligence and blockchain.

What I.AM brings to people, society, industry and the planet

The citizens/consumers have access to the service that they need when they want, wherever they are and with the equipment they have. In this context, the role of the brand is very strategic. Its valorization will be the result of the quality of the interactions between the members within a community.

With the Internet of Augmented Me, the asymmetry of information becomes favorable to the citizen/consumer who acquires bargaining power. Tools such as blogs, forums and recommendation sites give the viewer access to more information, especially the buying experience of citizens/consumers.

More generally, the objective is to smooth the exchanges, increase the ROCE of each company and allow it to have an apparent financial surface much larger than it really is.

The business model of I.AM

The characteristics of the business model of the Internet of Augmented Me are very different from traditional models that are essentially based on anything that can help push the product. With the Internet of Augmented Me, it is the opposite. The business model is based on the response to the latent expectations of consumer citizens and forecasting intelligence.

The PER of an "Internet of Augmented Me" enterprise is not that of an industrial company watch.

We go from the profitability of a company based on the MATERIAL capital to a company based on the IMMATERIAL capital.

The profitability of the Internet of Augmented Me enterprises is not that seen by the rational actors. It is measured by the ability to extract value from the QUALITY of interactions with citizens/consumers to create CONFIDENCE and therefore the ability to create a TRUST capital.

Box 1.7. *Takeaways*

1.6. Glossary

ATAWAD (AnyTime, AnyWhere, Any Device): this gives the possibility of accessing all or part of an information system anytime, anywhere and with any connected device, which implies that we can connect to the information system (IS) 24/7 without any interruption of access where we want it connected or not connected and especially with the equipment of our choice, especially with the seven screens (PC, smartphone, smartwatch, bezel, mirror, TV and tablet).

BaaG (Business as a Game): this summarizes the new way of doing business. The arrival of the Internet has led to a new way of exchanging information. Each anonymous participant has transparent access to all the information needed for the exchanges with a consensual process. No intermediary is involved in this process. BaaG triggers the "gaming" fiber to do business, which is the central engine of human motivation. Nowadays, when consumers expect high-quality services, speed and proximity, and when we are surrounded by a huge amount of data to use with care, BaaG will allow fluid interactions and massive decentralization.

PCE (Productivity Collaborative Exchanges): this measures and harmonizes how much communities deliver a concrete outcome based on the collaboration of its participants with its ability to connect, coordinate, process and categorize the information necessary for each citizen on a large scale. The problems are so complex and vast that they cannot be solved by any centralized organization. They must be solved in a decentralized way that results from the increased productivity of collaborative exchanges within each community. This unification is based on relations, attitudes and a culture of collaboration whose objective is to promote the common good. Business performance will depend on the quality and effectiveness of their services and the way they interact with their customers in communities.

TaaS (Trust as a Service): this is much more than security management. Trust is the key enabler for transferring its assets, processes or applications to a supplier. The bank is an excellent example of TaaS. If we put our money in banks, it is because they are synonymous with trust as a service (accounts receivable management, etc.). TaaS will be increasingly associated with its wallets and the ability to automatically secure and associate any transaction with a virtual trusted third party that will guarantee identification, money transfer, proof, autonomy, contract and traceability. These constitute the six letters of IMPACT.

Unimedia (unification of all media on the same platform): this is a term designed to gather several previously separate media together on a single medium. The new feature of the Unimedia is to offer contents that are fluid, malleable, protean, reproducible and transmissible in a network. They allow browsing and remote group work. The goal is to enable users to have unified data management and an interface that moves seamlessly from one medium to another. Unimedia must connect citizens/consumers in a transparent way, no matter where citizens/consumers are.

2

The Patterns of the Internet of Augmented Me

"Look deep into nature, and then you will understand everything better."

Albert Einstein

2.1. Why patterns?

The Internet of Augmented Me (I.AM) defines a living environment (a "universe") where citizen consumers are in control of their choices and directly influence the production of goods and services through electronic interactions, wherever they are.

This universe responds to the primary needs of individuals and their desire to flourish. It is based on combinations of advanced technologies, combining digital, quantum physics and biology.

The foundations of this universe are principles that guide the creation of solutions to the problems faced by individuals.

In 1977, the architect Christopher Alexander gave an answer with his pattern concept [ALE 77]: "Each pattern describes a problem which occurs over and over again in our environment, and then describes the core of the solution to that problem, in such a way that you can use that solution a million times over, without ever doing it the same way twice." We will describe each pattern with the following formalism:

– *description of the problem* to be solved;

– *description of the answer*: the elements of the solution with their mutual relations. The solution is called a *design template*;

– *description of the benefits for the consumer*, in line with the Internet of Augmented Me.

We identified 10 patterns articulating the universe of the Internet of Augmented Me:

– *zero everything economy*;

– *shazamization*[1] *of everything*;

– *nature, source of inspiration*;

– *Business as a Game*;

– *consumerization of everything*;

– *disappearance of technology*;

– *BOTization* and "*Build to Order for Me*";

– *virtualization of everything*, the era of *digital twins*;

– *decentralization of everything*;

– *automation of everything*.

2.2. "Zero everything economy"

2.2.1. *The problem*

Humanity is faced with profound transformations impacting our environment and life style. Those transformations are often induced by human activities. Our survival as a species could be compromised by:

1 Shazam is an app that identifies pieces of music (https://www.shazam.com/fr).

Climate change:

Global warming disrupts agriculture as well as vegetation and animal ecosystems. Melting ice and rising sea levels are threatening human life in some territories. In the 21st Century, humankind produces more and more greenhouse gases, CO_2 in industry and transport, methane with the expansion of livestock farming, and hydrofluorocarbons with refrigerators and air conditioners. These gases absorb infrared radiation from the Earth's surface and warm the planet.

Energy transformations:

– The overexploitation of the planet's resources, fish and forests, is endangering our way of life.

– Humans consume more resources each year than nature can replenish in a year. Humanity is living on credit. One day, called Earth Overshoot Day, symbolizes this shift towards more debt. In 2019, it was July 29. Twenty years earlier, it was September 29.

How can we ensure human well-being in a sustainable way?

2.2.2. *The solution*

The *zero everything* pattern provides a solution by offering to the *citizen consumer a simple and clear perspective, eliminating lame compromises.*

This involves identifying the causes of a problem and removing it.

For instance, chronic heat waves are driving up sales of air conditioners. Air conditioners that, on the one hand, cool down houses but, on the other hand, heat up the outside air. Creating more needs for air conditioners and electric consumption peaks compounds the initial problem. It is a vicious cycle.

In 2018, according to the International Energy Agency, air conditioners and electric fans already consumed a fifth of the electricity used in all the houses of the world, corresponding to 10% of the total consumption of electricity.

Air conditioning is partly responsible of the phenomenon of "urban overheating", by increasing the cities' temperature by 1–1.5°C compared with the surrounding countryside.

Air conditioning is without a doubt the wrong answer to the problem of cooling homes.

A zero air heating or even zero electricity solution is required.

We already have at our disposal a natural phenomenon to cool anybody. Radiative cooling at night emits infrared waves into far and cold space (at –270°C). Buildings as well as living beings reflect heat during this period and cool down. Yet, the sun remains an issue during the day.

Researchers at Stanford and Buffalo Universities have developed materials that reflect sunlight while absorbing ambient warm air and reflecting heat as a focused wave in the direction of deep space. This technique lowers the temperature of the building by a maximum of 6°C, compared to the surrounding air[2].

The mastery of nanophotonics[3] is the technological keystone of these systems.

However, the result is still insufficient during major heat waves, above 35°C. A solar air conditioning system with a closed circuit of hot and cold water becomes necessary.

If you also want the system to operate at night in heating mode, thermal storage is necessary.

The French company Helioclim[4] provides such a system in the form of an absorption machine using water and ammonia. The system is reversible and produces either industrial-grade cold, or domestic hot water at 60°C.

The keys to success lie in the technological ability to concentrate or reflect ambient heat back into space, all on a fixed surface. A bonus is to

2 See https://www.osa-opn.org/home/articles/volume_30/november_2019/features/radiative_cooling_harvesting_the_coldness_of_the_u/.

3 https://en.wikipedia.org/wiki/Nanophotonics.

4 http://en.helioclim.fr/.

produce electricity from solar radiation, therefore to use a photovoltaic system.

More generally, environmental issues are broken down into “net zero” objectives.

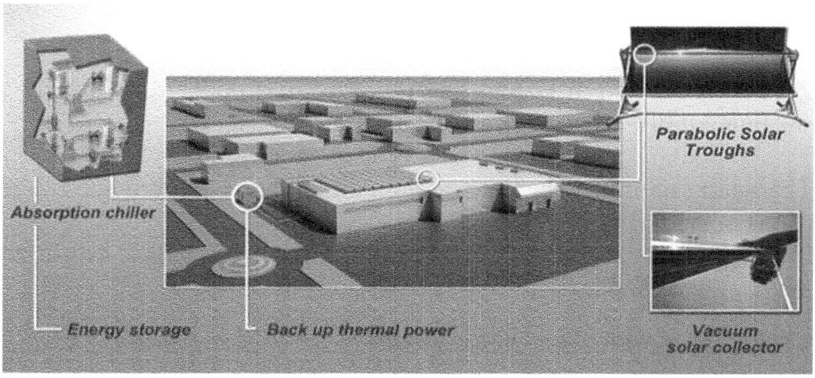

Figure 2.1. *Helioclim's reversible air conditioning system. For a color version of this figure, see www.iste.co.uk/duvaut/IAM.zip*

In September 2019, Amazon announced the objective of “zero fossil energy” for 2030 and “zero net carbon”, i.e. CO_2 neutrality, for 2040, for all its business operations.

Amazon's plan is based on electric mobility, and Amazon has ordered 100,000 delivery trucks from Rivian[5] in which the Internet giant has invested $500 million.

More ambitious in mobility is the plan of Mary Barra, CEO of General Motors, who announced her vision at the end of 2017:

– zero accidents: 35,000 Americans die every year on the road;

– zero pollutant emissions: pollution affects the development of the lungs and immune system of children, increases the birth rate of premature babies;

– zero congestion: the time spent in transport is 12 days per year for an American, which represents a loss of $305 billion for the American economy.

5 https://rivian.com/.

It is easier to understand why car manufacturers are playing the card of the all-electric and smart city.

Ford wants to create a new operating system for the smart cities and their lingua franca[6], and believes in the virtue of an urban infrastructure connected in 5G or by other C-V2X technologies (cellular communication vehicle for everything).

Zero waste is also in vogue. The vision of seas polluted by plastics and decimating wildlife has awakened Western consciousness.

The retail sector has begun to address the problem and provides the first solutions in the form of:

– biodegradable plastics;

– zero plastic with glass and cardboard;

– bulk goods.

Figure 2.2. *Zero waste trend in Franprix. For a color version of this figure, see www.iste.co.uk/duvaut/IAM.zip*

6 See https://interestingengineering.com/fords-smart-city-will-connect-everything-to-create-a-technological-utopia.

The French brand Franprix is innovative in this field. It has a *zero waste meal offer* for urban workers in a store in the center of Paris. Customers who use the roaster or the juice machine in the catering bar, managed by the food manufacturer Fleury Michon, can choose between disposable plastic packaging and glass containers whether they consume on site or take away their meals. In the second case, they can return the dishes later, so that they can be washed and then reused by the store. The price does not change in this formula.

Moreover, single-use plastic products such as straws and cups will be banned in the European Union by 2021.

Regulations associated with societal pressure will accelerate the development of ecological alternatives in the form of new bio-sourced materials, simple-to-operate recycling machines, and behavioral changes.

Synthetic biology and natural resources such as algae and operational waste will be heavily used.

2.2.3. *Consumer advantages*

The "zero everything economy" will offer consumers a healthier, safer, and more pleasant world. It is a lifestyle inspired by the natural functioning that allows ecosystems to live together harmoniously, making the best use of the territory's resources. This pattern is at the heart of an art of living in phase with human aspirations and respects the functioning of the living.

2.3. "Shazamization of everything"

2.3.1. *The problem*

For the consumer, choosing a product or a service is not a simple act when essential information is not available at the place of consumption.

This can lead to a dangerous situation: think of the health consequences of a food error for someone who has an allergy. An anaphylactic shock can be fatal.

Recognizing the nature of the food and its composition becomes a necessity.

In a dual way, simple and personalized access to services requires the identification of individuals in context, often coupled with an authentication process.

How can this information be obtained without explicit signage such as descriptive elements?

2.3.2. *The solution*

The word *shazamization* comes from "Shazam", the name of the application available on iPhone that allows you to find the title of a song from listening to an extract.

Its founder, an English company created in 1999, was acquired by Apple in December 2017. The algorithm has been integrated into Apple's voice assistant, Siri.

Shazamization now means the extraction of structured information from a physical object or real scene, such as its description and characteristics, from one or more sensors providing unstructured data.

For example, the Consumer Physics SCIO near-infrared microspectrometer allows us to scan objects and materials in our environment and recover their chemical composition on our smartphone[7].

It is the shazamization of food that makes possible the Dietsensor nutritional coach, essential for diabetics and for people with allergies to certain foods, such as peanuts, for example[8].

This service guarantees a healthy and balanced diet.

A variant of this system, called X-Spect, developed by the German company BSH and the Finnish company Spectral Engines, makes it possible to recognize each type of fabric and washing task. It transmits this information to a washing machine that determines the best washing program[9].

---

7 See https://www.consumerphysics.com/.

8 See https://www.dietsensor.com/.

9 See https://www.engadget.com/2017/09/02/bosch-x-spect-scanner.

Thanks to the shazamization of waste, the Bin-E[10] intelligent bin guarantees efficient recycling, simplifying the long and sometimes painful process of sorting for consumers. Ecology is becoming accessible to all.

We find the shazamization of individuals in biometrics, i.e. the use of biological characteristics such as the face, the iris and the voice for identification. The biological signature must be unique.

Biometrics is the key to access services. The object only works with its owner or with those it recognizes. Safety is a priority here.

For example, Apple's 3D face ID facial recognition system is used to unlock the iPhone X and to secure purchases while on the move.

For long distances, up to 200 m, and to compensate for poor light conditions, the US army has developed an infrared laser, capable of remotely detecting surface movements of an individual caused by their heartbeat.

2.3.3. *Consumer advantages*

Coupled with AI and a worn interaction object, such as an earpiece or glasses, shazamization increases individuals' senses and provides them with new perceptual abilities. We are talking about *augmented reality*.

Shazamization means contextualizing, allowing the realization of a physical Internet, where the consumer can make their decisions immediately, in full knowledge of the facts.

A new business will emerge, based on our daily experience and widespread shazamization.

We like a pair of shoes, worn in the street by an athlete doing exercises. With a camera and Snapchat's Visual Search function, you can find a similar product on the Amazon platform[11].

In short, shazamization allows you to build your style by imitating others, in a simple and instantaneous way.

10 http://bine.world/.

11 See https://techcrunch.com/2018/09/24/snapchat-amazon-visual-search/.

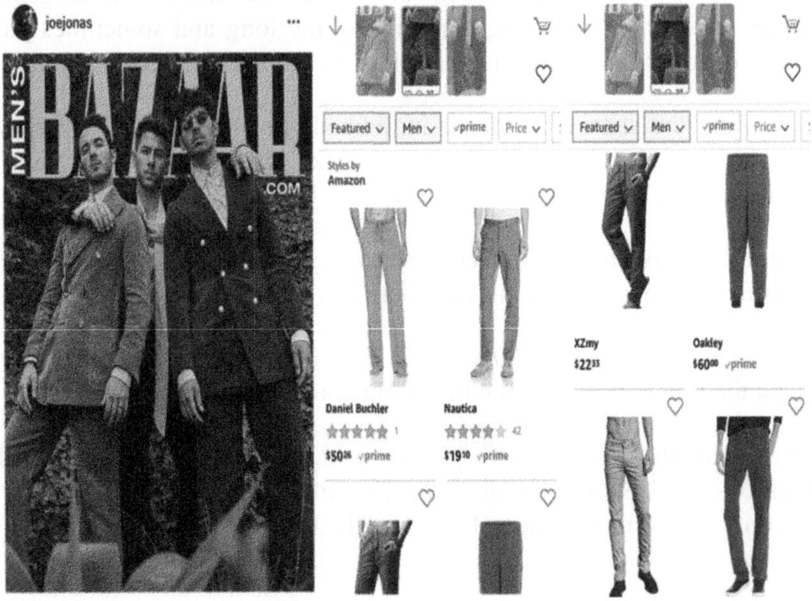

Not quite a match. *Photograph: Sarah Halzack/Bloomberg*

Figure 2.3. *Stylesnap. For a color version of this figure, see www.iste.co.uk/duvaut/IAM.zip*

2.4. Consumerization of everything

2.4.1. *The problem*

Health is becoming more and more technological. New medical imaging techniques (MRI, 3D scanners) require a lot of capital and expertise to analyze the results produced. In France, because of this cost, the number of these medical devices is limited via a health card and reserved for centralized technical platforms, promoting the extension of medical deserts and creating inequalities in access to healthcare in the territories.

How do we bring the service offer closer to the consumer? How do we make the service accessible from anywhere at a reasonable cost?

2.4.2. *The solution*

Consumerization is the process that transforms goods and services that are only accessible in B2B into goods that are accessible by consumers or consumer communities by using technological breakthroughs in consumer electronics.

The case of MRI is emblematic. This examination, which is essential to see the soft tissues of the body, requires a device with a powerful magnet. The more powerful the magnet, the more precise the image. The record is held by a 130-ton superconducting magnet delivering a magnetic field of 11.7 Teslas, 100 times more accurate than current imagers.

It is clear that such technology is still a high-tech, high-precision craft, to be undertaken by specialized technicians. This process is costly and it is difficult to envisage a reduction in system costs under these conditions.

Mary Lou Jepsen, CEO of Openwater[12], is using the electronic expertise she gained in the development of Pixel Qi displays to build low-cost, portable medical imaging devices with performance similar to a 2-ton MRI device.

Consumerization also means consumer involvement, and allows for DIY (Do It Yourself) practices. For example, the Neutrogenia brand presented a custom-made beauty mask service at CES 2019, based on the results of its SkinScan skin scanning smartphone accessory coupled with the Skin 360 app. The Esthetician's Pro quality arrives at the consumer level.

2.4.3. *Consumers' advantages*

The consumerization of everything makes it easier for the consumer to access services that were previously rationed or unaffordable. The consumer becomes independent and can more easily benefit from technological advances and acquire greater well-being.

12 https://www.openwater.cc/.

2.5. Business as a Game

2.5.1. *The problem*

To promote environmentally responsible behaviors, it is necessary to reconcile personal interest and the common good. This requires involving the consumers in their consumer choices, allowing them to monitor the impact of their efforts and to compare themselves to other individuals in the community.

How can such a virtuous cycle be set in motion?

2.5.2. *The solution*

Gamification of everything consists of transforming a commercial proposal into a game. As humans like to play to learn and for entertainment, these predispositions must be used to increase the use of applications and services and develop new skills.

The mechanics used are based on psychological springs such as emotions, challenges, improvement, progression, and self-accomplishment. For a team, it is about promoting social cohesion and group cohesion.

A form of regulation of the common good through play is a *nudge*. A nudge aims to influence people's behavior by using their cognitive biases, such as conforming to the opinion of the greatest number even if it goes against their convictions. Like an optical illusion, a nudge is supposed to have the same effect on those who are subjected to it. No constraints, no obligations, it is the individual who has the impression of making the decision themselves. For example, a hotel wishing to encourage its guests to reuse towels prefers to focus on the proportion of guests who are committed to the ecological cause. Being inexpensive to implement, nudges are easy to test.

Another technique, called the *Tamagotchi effect*, is based on the affection, the attachment that an individual in charge of a pet, can have and whose decisions will affect the life of said animal. As a Tamagotchi is a virtual animal, its success shows that a robot can be accepted in a home.

The *Eco Ego* children's game plays on these emotions by offering a series of questions about everyday actions, such as putting your shopping in a plastic bag or cardboard bag, and going to the store by bike or by car. The answers impact the life of an artificial nature, with penguins on an ice floe, trees in bloom or not. It is a way to teach children about eco-actions, which will then influence their parents.

A company's business choices and their impact on the market can be represented in the form of a game, a multi-agent simulation, where all players are virtualized as digital twins and an AI plays the role of a game master.

Finally, a company's business can be a game. Take the case of a company named Weward[13] which provides an app converting our steps into a "ward", a blockchain crypto-money that can be exchanged into coupons or euros. You win 1 "ward" after 1,000 steps, or two cents of euros. But you can earn more by visiting and consuming in certain shops. Weward thus improves the well-being of individuals while generating additional sales at partner merchants.

Blockchain technology is based on cryptography and game theory. It allows, via crypto-money dedicated to rewarding the good behaviors of those who go in the direction of the common good. In the case of Weward, it is a matter of fighting sedentary lifestyles and improving the health of populations, without any expense to the community.

However, this game values more collective actions, because only aggregated and cross-referenced data have value. French researcher Alain Rallet, in his text on "data valuation, 3 scenarios"[14] explains that "in the context of co-ownership of a building, or at the scale of a district or a city, individuals can decide to share some of their personal data such as their electricity or heating consumption, their mobility and environmental data, purchases, leisure, etc. to negotiate services using these data."[15]

Similarly, the Embleema[16] company, thanks to blockchain technology, brings transparency to the trade of patient data, by creating a decentralized market without intermediaries to allow each patient to exercise sovereignty

13 See https://www.weward.fr/.

14 https://ideas.repec.org/p/hal/wpaper/hal-01909650.html.

15 See https://hal.archives-ouvertes.fr/hal-01909650/document.

16 See https://www.embleema.com.

over their data. The more patients are involved in medical research, the more cryptocurrencies they receive.

2.5.3. *Consumer advantages*

When business becomes a game with clear and transparent rules, the consumer can then get involved and reap the benefits through rewards whose aggregate sum measures the amount of effort made for the common good. The consumer thus increases their self-esteem and receives recognition from the members of their community. The emotional benefit increases.

2.6. Virtualization of everything, the era of digital twins

2.6.1. *The problem*

Technological systems are becoming more and more complex, and their management requires advanced skills that are inaccessible to consumers due to lack of adequate training.

Managing a building or a factory in real time therefore requires a greater number of reporting and specialized staff. In the building industry, it is the job of facility managers to take on this task by taking care of the smooth running of the boiler room, elevators and lighting. The owner loses all control over the equipment in the building and its operation. During a renovation, it is necessary to re-map your technological environment by doing archeology. Many hospitals in France are aware of this situation.

Similarly, creating a machine learning model that can help a bar manager anticipate coffee replenishments requires the help of a data scientist, capable of manipulating the data and selecting the right algorithms. But this skill has a cost and the shortage of current experts makes the exercise inaccessible to small businesses.

In both cases, how can we allow a professional expert to take control of this increasingly technological world?

2.6.2. *The solution*

Virtualization of everything is the process that builds abstraction layers that can describe a hardware or software system.

In the IT field, virtualization is everywhere. The Unix OS in the 1970s described computer resources in file form: /dev/lp was a printer and /dev/cdrom was a cdrom drive. Partitions are also files.

Now, we virtualize operating systems to run them on host machines. We are talking about a virtual machine.

Two major 21st-Century technologies are based on virtualization: the Cloud and 5G networks.

The Cloud offers multiple services to the consumer by automating the provision of virtual machines, storage devices and the monitoring of their consumption in order to invoice the use of IT resources.

The 5G technology also relies on the virtualization of network functions or NFV (Network Functions Virtualization) to decouple control/command devices and the physical network infrastructure.

Without virtualization, there is no software-defined network, hence no possibility of centrally controlling the network and dynamically orchestrating its uses. This simplifies network administration and allows on-demand reconfigurations of network devices.

Virtualization of everything leads to the "*software-definted everything*" paradigm, in which the digital processing functions are transferred into a virtualized infrastructure.

A French company, Blade[17], is dematerializing the PC with its Shadow product.

Coupled with 5G, any display can access a PC with the appropriate resources (memory, CPU, GPU) to offer a quality service, with the end of programmed obsolescence in sight. This allows gamers to play the latest games without physically changing their computers.

---

17 https://www.blade-group.com/en/.

If you can virtualize a computer and a computer network, you can virtualize any object as soon as it is connected. The digital twin is the result.

A *digital twin* is the virtual representation of a product used for product design, simulation, monitoring, optimization or maintenance. This concept is at the heart of the "Industry 4.0" program and the intelligent building of information systems.

In the building industry, virtualization is merged with digital modeling of the building, equipment and sensors/actuators. The BIM (*Building Information Model*) standard is a structured set of information describing these objects, their characteristics and the relationships between them. The materials of a wall, the location of a boiler, and the geometric description of a part are parts of the BIM modeling. This goes beyond the geometric description of the building provided by architectural design software.

2.6.3. *Consumer advantages*

Virtualization and the Internet of Things make cockpit and dashboard applications possible and allow the individual to understand a situation in the blink of an eye.

Understanding to better act on the system, and the contribution of 5G with its latency in the millisecond range is essential for real-time remote operation.

Virtualizing your environment is the first step towards controlling it.

2.7. The technology disappears

2.7.1. *The problem*

Consumers do not buy technology anymore, they buy an experience that uses technology. In other words, you do not buy an OLED 4K HDR ultra-flat TV, you buy the possibility of watching a football match with your friends over a beer.

Marc Weiser wrote in 1991: "The deepest technologies are the ones that disappear. They blend into the flow of daily life until they become indistinguishable from it." [WEI 91]

How do we make technology both invisible in our environment and available at all times?

2.7.2. *The solution*

Let us take the case of the computer.

As part of the European "Disappearing Computer" program, which included 17 projects from January 2001 to December 2003, Norbert Streitz, founder of the Smart Future initiative[18], a source of new ideas and expertise on technological trends that impact our lives, defined two types of technological disappearance.

(1) The *physical disappearance* allowed by the extreme miniaturization of electronics, which favors integration onto our body by being grafted onto worn objects such as watches or clothing or into the environment.

(2) The *mental disappearance* makes the computer invisible to our eyes by blending it into the scenery, a wall, or furniture.

Further, this disappearance can be *either cognitive or emotional.*

In the first case, the appearance of the electronic object is changed and the user no longer perceives it as such. This is the case with Samsung's ultra-flat QLED TV "The Frame", hung on a wall displaying a work of art when the TV is turned off. The user sees a classic painting with its frame, except that the work presented may change.

Similarly, furniture manufacturer Ikea, in partnership with Sonos, offers a WiFi speaker connected in the form of an auxiliary lamp, Symfonisk. Here again, the user sees a lamp capable of diffusing sounds on demand, and not a classic speaker with a rectangular or cylindrical shape.

In the emotional case, the shape, texture or visual appearance of the object, handled or not, induces in the user an intense psychological

18 https://www.smart-future.net/.

experience. The Nabaztag rabbit[19], proposed by the French company Violet in 2005, falls into this category. The rabbit's ear movements are supposed to reflect its emotional state and capture our full attention. We are in the middle of sharing experiences from a distance.

The French company Hayo offers a 3D sensor in the form of a 22-centimeter-high black cylinder capable of interpreting the user's movements at specific locations and then activating a programmed function accordingly. Any object in the scanned room can become a remote control of our environment and activate behaviors in given postures. An almost magical dimension emerges and provokes positive sensations in the user.

2.7.3. *Consumer advantages*

To make technology disappear is to restore the full meaning of each individual's lifestyle and give her/him greater power over her/his environment without transforming her/him into a cyborg.

Consumers can therefore choose the universe in which they wish to live, while benefitting from technological comfort.

2.8. Nature as a source of inspiration

2.8.1. *The problem*

Digital technologies used by humans are increasingly energy consuming. If the Internet were a country, it would be the third largest electricity consumer in the world, with 1,500 TWH per year, behind China and the United States. In total, digital technology consumes 10–15% of the world's electricity, the equivalent of 100 nuclear reactors. And this consumption doubles every four years: this is hardly acceptable at a time of climate change and when we need CO_2-neutral approaches.

Current technologies use rare metals, such as lanthanum, indium or iridium. The most advanced technologies such as electric batteries, fuel cells, electronic circuits, optical fiber, wind turbines and smartphones make it a major consumer. The rarity comes from the fact that they are present in

19 https://www.nabaztag.com/.

small quantities in the Earth's crust and that their exploitation requires carrying tons of soil, and using polluting chemicals that bring multiple diseases to the surrounding populations. The exploitation of these mines creates environmental disasters.

In the Congo, cobalt is a source of conflict between warlords, because its exploitation is a source of income.

China, which controls 95% of rare metals production, makes it an instrument of blackmail to access Western technologies.

Guillaume Pitron, in his book *La guerre des métaux rares*[20] (published as *The Rare Metals War* in English) shows that the current energy transition, which is based on electric vehicles and ever more on digital networks, is a decoy. He states [PIT 18] that "Over its entire life cycle, an electric vehicle generates almost as much carbon as a diesel."

How can we ensure that technological progress, which has enabled billions of people to escape poverty, does not destroy the environment and lead humanity to a destructive impasse?

2.8.2. *The solution*

Biologist Janine Benyus proposed in the early 1990s a new approach in her book *Biomimicry: Innovation Inspired by Nature* [BEN 97]. According to her, nature gives us lessons:

– using one main energy source: *solar energy*;

– using only the amount of energy we need;

– adapting the shape to the function;

– recycling everything;

– rewarding cooperation (including *symbiosis* and *sustainable interactions*);

– betting on *biodiversity*;

---

20 https://fr.shopping.rakuten.com/mfp/5910902/la-guerre-des-metaux-rares-la-face-cachee-de-la-transition-energetique-et-numerique.

– requiring local expertise (humans, *chimpanzees*, *parrots* and *sheep* are able to fulfill needs in their environment, when needed);

– limiting excesses from the inside;

– using constraints as a source of *creativity*.

In other words, nature, with its 4 billion years of R&D, is an unrivalled source of inspiration for finding solutions to these problems that threaten humanity's survival.

The human being is the most successful product of this evolution.

We are mainly composed of oxygen, carbon and hydrogen.

Carbon is a material that is abundant in nature. It is the building material of plants that absorb CO_2 every day, extract carbon by photosynthesis and release oxygen. Forests are carbon neutral.

To enable carbon-neutral human activity, new carbon cycles must be created. The best way is to test these solutions on a multi-activity industrial zone, trade and industry. An airport or a port is an ideal playground.

Air transport accounts for 2–3% of global CO_2 emissions. Airports, as objects of vital interest, must be energy self-sufficient to operate in all circumstances.

A 100-passenger electric aircraft capable of flying 500 km in Europe is scheduled for 2030, as current batteries do not provide enough energy to run powerful engines over long distances[21].

In addition, the switch to electric power requires the complete replacement of fleets, which is economically inconceivable at the moment.

A quick solution is to operate aircraft with bio-kerosene that does not require the immobilization of arable land needed for human food.

21 See http://www.leparisien.fr/high-tech/les-avions-hybrides-prennent-leur-envol-09-04-019-8045631.php.

In 2019, Rotterdam The Hague Airport began building a demonstration plant to produce 1000 liters per day of this fuel using CO_2 extracted from ambient air using the technology of the Swiss company Climeworks and hydrogen produced by electrolysis from water and electricity generated by solar panels[22].

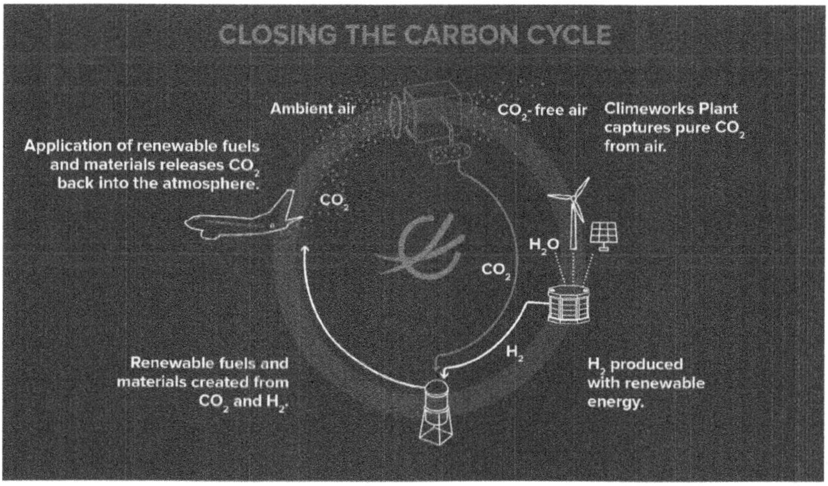

Figure 2.4. *Closing the carbon cycle. For a color version of this figure, see www.iste.co.uk/duvaut/IAM.zip*

Hydrogen can also power fuel cells in aircraft used to travel on the tarmac during takeoff or the landing phases without drawing on fuel reserves.

This represents a 4% saving on fuel consumed and a 7% reduction in CO_2 emissions[23].

The CO_2 and H_2 pairs can also be used to produce proteins that can be used for animal feed, beef or fish. As a result, agricultural land growing soybeans is freed up for livestock feed only. This is a form of decarbonation of agriculture[24].

---

22 See https://www.climeworks.com/wp-content/uploads/2019/08/Press-release_Rotterdam-The-Hague-Airport-initiates-study-for-the-production-of-renewable-jet-fuel-from-air.pdf and https://www.climeworks.com/renewable-jet-fuel-from-air/.

23 See https://www.mega-piles.com/news/easyjet-pile-a-combustible-707.

24 See https://bioenergyinternational.com/research-development/30229.

CO_2 in supercritical form, above its critical pressure and temperature, serves as a solvent for dry cleaning clothes, without the health disadvantages for workers in the sector.

To produce living matter or bio-fuel, CO_2 is an excellent material.

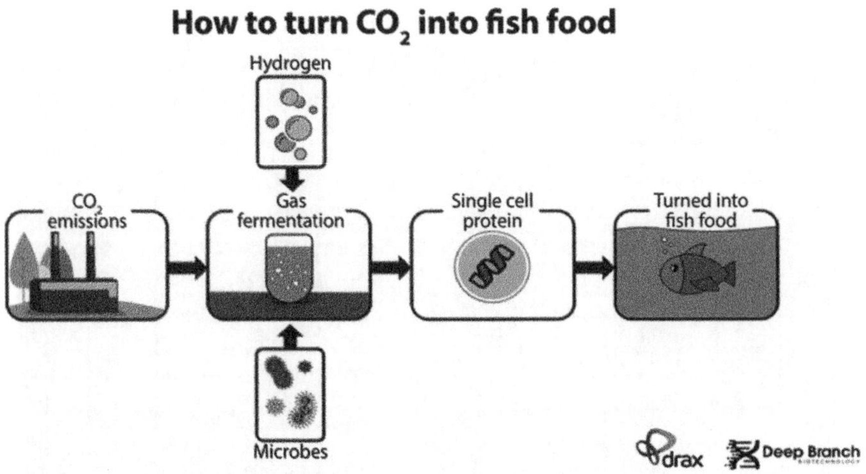

Figure 2.5. *How to turn CO_2 into fish food. For a color version of this figure, see www.iste.co.uk/duvaut/IAM.zip*

When it comes to producing, transporting and storing electricity, two forms of carbon dominate: graphene and carbon nanotubes.

Graphene, a monolayer of carbon honeycomb sheets, will be the star material in our digital world.

Without it, there would be no miniaturized high-gain 5G antennas capable of covering a long distance, no Skeleton[25] super capacitor capable of halving the consumption of an elevator and recovering braking energy during the descent.

Graphene also makes it possible to produce clothing that dissipates heat and moisture from the body, is antibacterial and conducts electricity, paving

25 https://www.skeletontech.com/.

the way for innovative uses, such as charging a smartphone in one's pocket[26].

The *carbon nanotube* is a rolled graphene sheet.

Here again, this material has extraordinary properties.

The US company Nanosolix[27] plans to use it to produce rectifier antennas, such as RFID tags that transform electromagnetic energy into direct current, capable of recovering up to 90% of the solar energy received as electricity.

Such performances will make it possible for solar cars to be completely autonomous.

Currently, on the Hyundai Sonata Hybrid, the solar panels on the roof provide 30–60% of the battery charge.

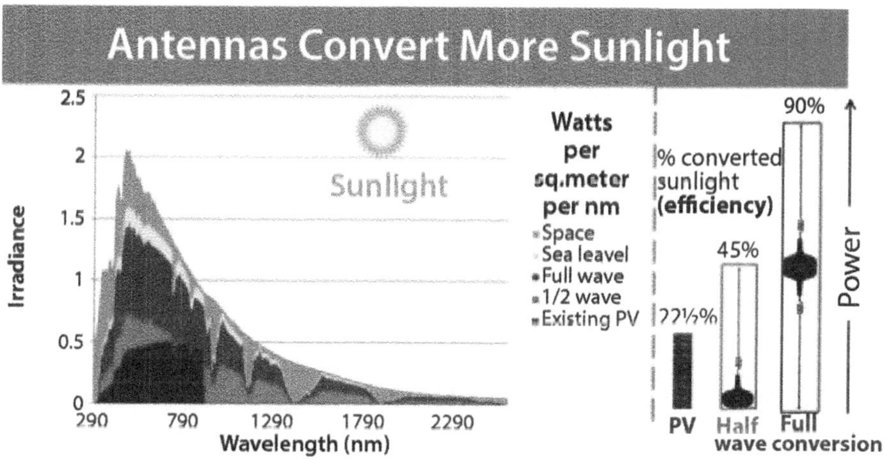

Figure 2.6. *Antennas convert more sunlight. For a color version of this figure, see www.iste.co.uk/duvaut/IAM.zip*

26 See https://www.fastcompany.com/90205090/the-first-graphene-jacket-is-here-and-its-magical, https://www.skeletontech.com/news/press-release-graphene-cuts-elevator-energy-consumption-in-half and http://www.newelectronics.co.uk/electronics-blogs/what-will-a-phone-look-like-in-ten-years/217432/.
27 http://www.novasolix.com/.

Carbon nanotube electrodes produced by the Nawatechnologies[28] company make it possible to produce batteries with high energy density and the capability of recharging in a few seconds.

Carbon nanotubes open the door to quantum computing, by allowing, coupled with bleach and ultraviolet light, to create unit sources of photons that could act as *qubits* (quantum bits)[29].

Finally, the combination of graphene and carbon nanotubes makes it possible to produce high-performance filters to desalinate seawater.

The most abundant energy in the universe is thermonuclear energy, which is found in our sun and brings us life.

A British company, First Light Fusion[30], has an original inertial confinement approach, inspired by the pistol shrimp that slams its claws hard to produce a shock wave that stuns its prey, and creates a cavitation phenomenon in the surrounding water. The steam and air cavities implode, generating hot plasma.

The company hopes to achieve an energy gain in 2024.

A human brain consumes 20 W and allows us to understand our environment in real time, to adapt and to solve complex problems. The AI Alphago crushed Go champion Lee Sedol. It consumes 1 MW, 50,000 times more energy than the human champion, and is unable to do anything other than play Go.

Recently, researchers at the University of Amherst, Massachusetts, calculated that training a common AI model using deep learning techniques produced as much carbon as five standard American cars, with the calculation taking into account the entire life cycle. In addition, the acquisition of new skills requires rebuilding the model from scratch. This coarse imitation of the neural system does not, and so far cannot, scale up.

28 http://www.nawatechnologies.com/.

29 See https://phys.org/news/2019-08-single-photon-source-household.html and https://phys.org/news/2019-06-carbon-nanotubes-graphene-based-membranes-desalination.html.

30 https://firstlightfusion.com/.

The architecture of the human brain provides an answer. Computing with electronic analog circuits mimicking neural systems is called "*neuromorphic computing*".

The French company Anotherbrain[31] is about to produce such a circuit, capable of continuous unsupervised learning with little data, at a low energy cost. Here again, nature inspires the solution.

2.8.3. *Consumer advantages*

Inspired by nature to design everyday products and services is the promise to benefit from the advantages of technology without destroying the environment, disrupting the climate and thus jeopardizing the very existence of humankind.

There is no Internet of Augmented Me without eco-responsibility and resilience.

2.9. BOTization of Everything, "Build to Order for Me"

2.9.1. *The problem*

The consumers of the *I.AM era* want highly personalized products and services at a reasonable cost. This means the implementation of a generalized mass customization.

Dell invented such a system in the 2000s for the sale of its computers on the Internet.

The consumer chose the computer components she/he wanted to integrate into her/his computer, paid by credit card and received the computer a few days later.

The BOTization of everything means extending this process to other industrial and service sectors: selling before producing.

31 https://anotherbrain.ai/.

2.9.2. *The solution*

BOTization is the industrial translation of the C2B2B2C model, *Customer to Bot to Bot to Customer*, at the heart of *I.AM.*

The current version of the *BOTization* involves a ubiquitous interaction pattern allowing the consumer to order an object or service of her/his choice everywhere she/he is.

The smartphone, which has replaced the PC as an interaction object for commerce, is such an object. But already, the post-smartphone era is coming, with connected objects allowing dialogue in natural language, such as the connected speakers Amazon Echo and Google Home.

But a smartphone can also act as a mirror, a radio alarm clock or a thermostat.

In a mobile situation, you can use the GPS connected in the car or glasses or earphones.

The interaction object is the entry point to a personal bot, an AI capable of understanding the consumer's expectation and determining the best way to meet it. To this end, the personal bot will contact a business bot via a marketplace to translate the demand into industrial or service reality. A *universal interbot dialect* is a necessity to achieve this objective.

In the factory, the bots are embodied in the form of robots, driving 3D printers that will be in charge of industrialization and object manufacturing.

Once this step is completed, the more or less robotized supply chain will deliver the object in question or the service to the person where the person wants it.

2.9.3. *Customer advantages*

The BOTization of everything makes it easy to give life to the lifestyle you desire.

This allows personalities to flourish and live a life that suits them.

2.10. Decentralization of everything

2.10.1. *The problem*

The consumer wants to be increasingly involved in the manufacture of her/his environment and the objects that make up her/his universe. There is a strong desire for autonomy from a distant and centralized industrial world that often imposes its rules and limits the choices of consumers, not responding to their values.

Until now, consumers have expressed their values and their style through creative leisure activities, for decorative objects, or through DIY.

2.10.2. *The solution*

Decentralization of everything is reflected in the phenomenon of *DIY*, short circuits and P2P exchanges (peer-to-peer).

When it comes to DIY, the factory arrives at the consumers' premises. They produce their own beer, their own bread, their own coffee and their own consumables.

Thus, a major brand such as LG proposed to CES 2019 a machine to manufacture beer based on malt capsules, yeast, hop oil and perfumes. However, the capsules are disposable and do not fit into a "*zero waste*" policy.

A more virtuous example in the ecological sense is the coffee capsule maker Capsulier, seen at CES 2019, which allows you to make your own Nespresso coffee or tea capsules with the mixture that suits you best[32].

Reusable capsules avoid the material waste of used Nespresso capsules.

The DIY vogue blends with the biological wave, and a lot of web content explains how to make natural and effective home care products yourself, which greatly reduces the number of chemicals stored in cabinets.

---

32 See https://www.capsulier.com/.

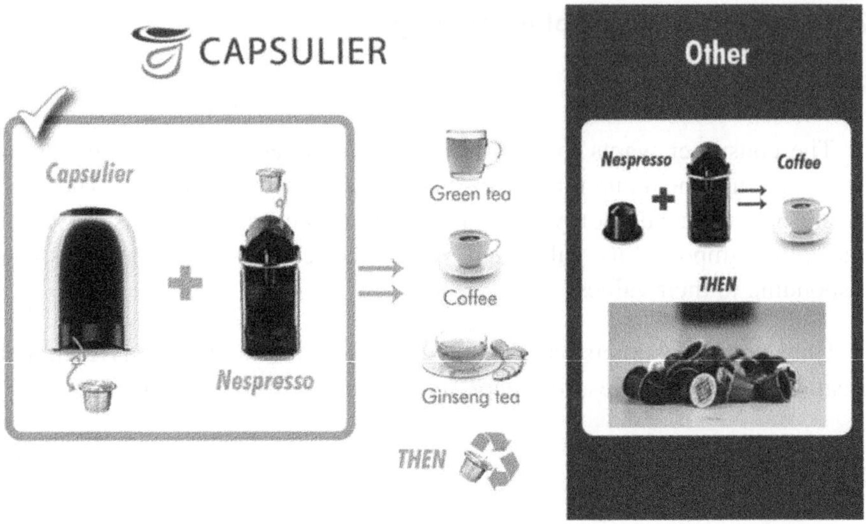

Figure 2.7. *Capsulier*

The phenomenon also affects beauty, with a proactive and more personalized approach.

For example, the French company BeautyMix[33] offers a small blender and accessories in order for the buyer to be able to make natural cosmetic products themselves.

More importantly, individuals can now produce their own cooking gas, their own electricity.

For example, the Israeli company Homebiogaz[34] makes it possible to transform food and human waste into cooking gas, and the Portuguese company BeON[35] offers the first "plug and play" solar kit, which plugs into an electrical outlet and provides partial power to the house.

Short circuits make it possible to connect the producer directly to the consumer, without going through traditional commercial intermediaries,

33 https://www.beautymix.fr/.

34 https://www.homebiogas.com.

35 http://www.beonenergy.com/.

wholesalers and retailers. This consumption pattern favors local producers and relies heavily on web technologies.

Short circuits reduce the number of intermediaries, but do not eliminate them. The consumer moves from a suffered intermediary, usually a monopoly platform, to a chosen intermediary.

Blockchain technology is particularly suitable for this type of trade, as it makes it possible to track trade linked to production and consumption. This makes it possible to create forgery-proof and transparent certificates of origin, a digital double of the registered designation of origin so dear to local producers.

Walmart, thanks to the Hyperledger Fabrik blockchain of the Linux Foundation, is able to trace the origin of 25 products from five different suppliers. The trader will soon ask all their vegetable suppliers to adopt this system in order to create operational transparency and better internal accounting.

In China, Walmart is working with the Singapore based company Vechain[36] to achieve even more ambitious goals.

It is possible to associate a cryptocurrency to a blockchain to reward intermediaries offering good advice. This creates a new distribution channel.

The French company BTU Protocol uses this mechanism to offer a hotel or car reservation with zero commission.

Finally, *P2P exchanges* of goods and services are the foundation of the collaborative economy and are carried out via commercial or non-profit platforms. It is about making the best use of capital: mobility capital through the best use of one's car, or energy capital through the monetization of the flexibilities generated by decentralized production (i.e. energy produced and not consumed by its owner).

The collaborative economy encourages the notion of access to the services provided by a good and not its ownership.

36 https://www.vechain.com/.

The place of exchange is often a marketplace, and blockchain technology via the distribution of *crypto-money* makes it possible to reward good behavior and those that promote the common good. These *crypto-money* systems can act as generalized coupons for the consumption of services in a virtuous ecosystem. For example, your energy trading earnings from your home can finance all or part of your electric mobility.

2.10.3. *Consumer advantages*

Decentralization of everything leads to more autonomy for consumers but also to more responsibilities. Consumers control their environment and contribute effectively to a sustainable economy by making the right choices. Local exchanges increase the richness of the territory and contribute to creating a social and civic bond within society.

2.11. Automation of everything

2.11.1. *The problem*

For the time being, the digital revolution has not eliminated repetitive and painful tasks for factory workers. These tasks are sources of errors and mistakes. For example, drilling and riveting by hand is very tiring for arms and knees. This causes musculoskeletal disorders in workers on these workstations and causes losses in accuracy, defects and therefore reduces the quality of the result.

At the same time, administrative processes often rely on the creation of files with data from several software programs that are unable to communicate with each other. Successive copy and paste operations are used to complete forms. These tasks have little interest and value, but repeatedly mobilize white-collar energy.

2.11.2. *The solution*

Automation is not a new phenomenon. It is at the heart of the industrial revolutions that have shaped our world, from James Watt's steam engine in 1769 to the assemblies of robots working in the new fuselage assembly line for the Airbus A321 aircraft in Hamburg.

In this plant, all the elements of a production line are automated, from its logistics and digital data collection system to its robots. A single person manages 20 robots and the entire production of the fuselage.

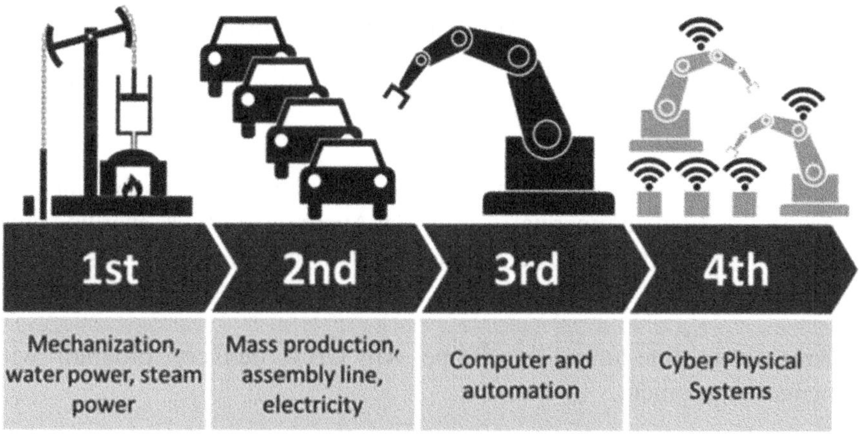

Figure 2.8. *Fourth industrial revolution*

At the same time, distribution chains are being automated, driven by two major players, Amazon and Ocado.

Ocado[37] gets the most out of warehouse space by orchestrating precisely the "ballet" of the robots that populate it. Ocado built the warehouse operating system based on a digital twin.

The delivery of parcels is also becoming automated, with the development of autonomous vehicles and drones. For example, Amazon has invested $500 million in the American company Aurora[38], which provides a software stack capable of making vehicles autonomous. Aurora is the Microsoft of the car industry.

Similarly, Ocado has partnered with Aurora's English competitor, Oxbotica[39].

During CES 2018, Toyota presented the e-Pallet model on stage. It is a modular autonomous vehicle concept designed to transport people or goods.

37 https://www.ocadosolutions.com/.
38 https://aurora.tech/.
39 https://www.oxbotica.com/.

The chassis is the same, but the bodywork and interior trim vary according to the uses.

Toyota has created an alliance to promote this concept, which already includes major names such as Amazon, Pizza Hut, Mazda, Uber and Didi.

The first e-Pallets are set to operate during the Tokyo 2020 Olympic Games.

UAV delivery is also expanding, and Alphabet's Wing subsidiary has launched the first commercial service in Australia. UPS is working with California-based Matternet to deliver medical equipment to hospitals in North Carolina.

Amazon is fine-tuning its Prime Air UAV delivery program with homemade equipment.

Hardware is not enough, one also needs software for autonomous control. It is also modular and based on an operating system in the form of a platform of weakly coupled micro-services.

Generally, such an operating system is open source because it allows you to quickly build a developer community and facilitates the creation of an ecosystem of value services.

EdgeXfoundry[40] is an example, adapted to the industry.

The more complex a system to manage, the more necessary it is to integrate a simulator into the operating system. The aim is to verify in the virtual world that the strategy implemented to achieve an objective does not present dangers or inefficiencies not previously perceived. Humans accomplish this task naturally with their prefrontal brain. A conductor has already played his score in his head before going into concert mode, as well as a champion who has repeated her race many times before.

Thanks to the wireless 4K virtual reality, the 5G networks and a haptic feedback allowed by an artificial skin, it is possible to make the virtual experience very close to reality.

40 https://www.edgexfoundry.org/.

When it comes to robots, the open-source operating system ROS (*Robotic Operating System*)[41], offers a simulation software, Gazebo[42]. This allows one, in complex and uncertain situations, to test different action plans before choosing the most appropriate one.

In warehouses, the automatic handling of unfamiliar objects in an unknown and unstructured environment is a real challenge. Amazon organized the "Pick and Place" challenge for three years before giving up in 2017. The machine was unable to reach the performance of a human picker in a warehouse.

The French company Conscience Robotics[43] offers a universal artificial intelligence capable of taking control of any type of robot: bipedal, multi-species, wheeled steering or even underwater. The robot learns by itself to take advantage of its physical capacities in order to move and act autonomously and optimally. The solutions are continuously improved and shared with all robots. The robot analyzes its environment, models its content in real time and shares it with all authorized robots.

Companionship between robots is becoming a reality, just like the Airbus companions who share their knowledge.

Automation also affects administrative tasks, with RPA (*Robotic Process Automation*).

Software robots automatically perform routine, rule-based processes.

Several business processes are involved, especially in finance, insurance and accounting, wherever there are repetitive entries. The following tasks can be automated:

– in payable accounts, invoice acquisition and approval, control execution;

– in the audit of multiple payments;

– in customer accounting, invoice generation and validation, payment reconciliations, dispute analysis and processing;

41 See https://www.ros.org/.

42 http://gazebosim.org/.

43 https://conscience-robotics.com/.

– in general accounting, complex accounting entries, account reconciliation, calculations of expense allocations, VAT returns or maintenance of asset records, payroll processes, management control and monitoring.

Uipath[44], Europe's leading RPA company, promises to offer distributors technology as effective as Amazon's in managing its platform.

In short, an Ocado/Uipath combination would allow JCPenney, or Macy's, to compete with Amazon on all its capabilities.

In short, the creation of Amazon OS becomes possible.

Automation is also affecting the home, with the appearance of specialized robots such as the LG Home-Bot robot vacuum cleaner or the Farmbot open source gardener robot[45]. Though currently, no robot can successfully perform the tasks of a housekeeper such as washing clothes, making beds, washing windows and cooking.

Toyota's research laboratory, the Toyota Research Institute, has decided to take up the challenge by allowing humans to train robots via virtual reality[46].

2.11.3. *Consumer advantages*

The citizen consumer is also a producer, and has a job either in a factory or in the world of services. The automation of everything allows them to escape the painful aspect of work and its disastrous effects on their health. The citizen can devote themselves to more noble tasks, highlighting their knowledge of being and their ability to manage complex situations. For the time being, we can see that greater robotization of companies, often linked to an influx of orders, does not lead to job losses.

For the consumer, it will be a question of moving from a time spent to a chosen time, of devoting their energy to activities that fascinate them: sport, music and creative leisure.

---

44 https://www.uipath.com/.

45 https://farm.bot/.

46 See https://futurism.com/the-byte/toyota-vr-teach-home-robot.

2.12. Patterns are linked together, they are design compasses of the Internet of Augmented Me

A pattern alone is usually not enough to define an I.AM solution. A combination of patterns is necessary. This requires an understanding of the coherence and the links that bind the patterns together, as explained below.

> *The zero everything economy* is the most important pattern, because it gives a direction to follow and defines clear objectives, which are structuring for the companies.
>
> *Decentralization of everything* emphasizes the local level and the responsibilities of individuals within a community. It indicates that everyone must be involved to achieve the objective.
>
> Without *consumerization*, enabled by advanced technologies using biology and electronics, it is difficult to give everyone the economic means to achieve the objectives set by communities and to meet the needs of individuals.
>
> *Nature*, because it finds energy-efficient solutions and relies on abundant resources, is a major source of inspiration.
>
> *BOTization of everything* coupled with *Business as a Game* makes it possible to define sustainable trade as long as the rules of the game are defined in a way that respects natural balances as well as the autonomy and values of individuals and communities.
>
> *Automation of everything* must free individuals, allowing them to choose new ways of occupying time. *Automation of everything* is based on *digital twins* brought by *virtualization of everything*.
>
> The Internet of Augmented Me must therefore have a legal component, a community regulation, or at the level of blocks of nations, to ensure compliance with the founding principles.
>
> The principle of European subsidiarity is an echo of the pattern of *decentralization of everything*.
>
> *Shazamization of everything* makes it possible to guide the choices of individuals in a given situation by providing them with the information they need, and to visualize in real time the efforts made for the common good via the game measurement instruments.
>
> The "*technology disappears*" pattern guides choices in the design of objects, products and services. It allows the citizen consumer to build her/his own lifestyle.

Box 2.1. *Links between patterns*

To paraphrase the great economist John Maynard Keynes, we must put the economy back at the service of society, at the service of an art of living, in which freedom, love, friendship and the arts have their place.

In the end, the patterns of the Internet of Augmented Me define this famous Art of Living dear to Keynes in terms of production, distribution and lifestyle. They meet the requirements of a sustainable economy without sacrificing the comfort and quality of life of individuals.

They help to choose the technological bases, modes of organization in society and economic models supporting the Internet of Augmented Me.

The patterns are the compasses of the architects of this universe.

3

Edge Computing for Processing and Connectivity in the World of Internet of Augmented Me

"A computer would deserve to be called intelligent if it could deceive a human into believing that it was human."

Alan Turing

3.1. Edge computing

3.1.1. *Summary*

When studying the entire chain of processing and connecting devices and servers for an end-to-end solution viewpoint, two main approaches emerge. The first one, intuitively, is "cloud first" or sometimes even "cloud only", where all the computing intelligence comes from cloud-based servers. The second is "mobile first", where local computing can provide services to the end-user "at the edge" of the Internet.

The second is what is meant by "edge computing". The aim of this chapter is to explain why we think edge computing is so important.

According to most surveys on the IoT (IoT being the generic term for local devices, sensors and other connected things), irrespective of the actual figures involved, the number of units to manage will increase in the coming years. As an example, consider the June 2019 forecast. The IDC estimates that there will be 41.6 billion connected IoT devices, or "things", generating 79.4 zettabytes (ZB) of data in 2025[1].

How can we deal with this, adopting only a cloud-centric view? Is it really possible to connect all these devices, sensors, etc., to cloud services and to consume these services only over the Internet?

That is the question edge computing aims to answer.

EDGE COMPUTING IN A FEW WORDS.– Edge computing represents all tasks performed as close as possible to the location of data creation or end-user usage. While it complements them, it is the opposite of all the tasks performed centrally, at the cloud level.

> According to Matt Vasey, who focused on IoT strategy at Microsoft, The ideal use cases [for both fog and edge computing] require intelligence near the edge where ultra-low latency is critical, run in geographically dispersed areas where connectivity can be irregular, or create terabytes of data that are not practical to stream to the cloud and back.

3.1.2. *Edge computing is a fundamental non-visible part of online services*

EDGE IS A NON-VISIBLE ASPECT NEEDED TO DELIVER BEST UX (USER EXPERIENCE).– In the real world, users do not really know about logistics, warehousing, etc., when they receive a package by mail. Edge computing is, in comparison, the same non-visible part of the UX. Users do not see it and startups spend a lot of time, effort and money on adapting and managing it.

When talking about edge computing in the digital world, we can take the view that it is a matter of processes (also called computing), data (from storage to usage) and networks (from network connection to data transmission). These three aspects will be used to discuss edge computing in the rest of this chapter. This creates a complete and mobile digital experience for end-users, in both personal and professional areas.

1 DC – https://www.idc.com/getdoc.jsp?containerId=prUS45213219.

This chapter is centered on the non-visible part (for end-users) of digital services used on a day-to-day basis. Do you use a smart phone? A smart watch? A connected object? A drone? None of these devices existed 15 years ago and they are now useful and valuable. To deliver the services required, technology firms in the edge marketplace have built a "stack of technologies" as a powerful and dynamic ecosystem of tools. This stack lives in a complex and complete integrated marketplace from the viewpoints of the back office, "the cloud", and the front-end, where the digital experience is delivered with a combination of intervening technologies.

Even though end-users do not see it, this stack of technologies is the basis for entrepreneurs involved in the digital world. If startup managers ever forget the complexity and density of this technical world they have to master, they will fail. If you are a startup manager, that is certainly why your CTO is your best friend and why investors want to know exactly who they are.

As a comparison, in a real-world business, you cannot ignore the impact of basics such as the supply chain, warehouse management, transportation, industrial processes, etc., on your success or your failure. All of these non-visible parts of your business, which your customers do not see, are important and can become critical whenever they fail. They are directly linked to the quality of your service, and if they fail, your company will not deliver the service it promises.

This is the position for edge computing in the digital world: if the edge does not work, the service does not work.

3.1.3. *From the past to now, edge computing is a natural evolution*

Edge computing comes with the growing maturity of the Internet, as connected PCs followed the maturity of mainframes many years ago. It is the natural cycle of innovation: when central computing reaches maturity, local devices extend central capabilities, and when local computing in turn becomes mature, central computing integrates, pools or standardizes local capabilities and expands its own capacity. This is shown in Figure 3.1.

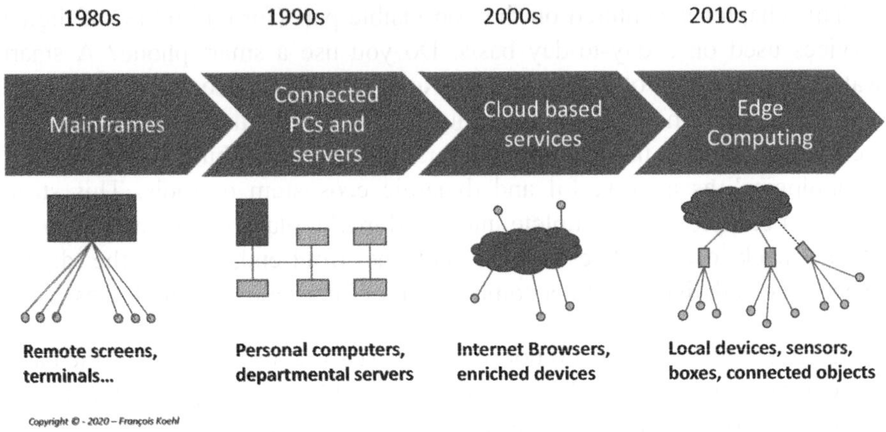

Figure 3.1. *The evolution of computing. For a color version of this figure, see www.iste.co.uk/duvaut/IAM.zip*

3.1.4. *Do you need a killer service to demonstrate edge computing's reality and benefits?*

Everything imagined, designed and built around autonomous vehicles creates an impetus to build a new world where local sensors, devices and usages would form the main part of a global solution for integrating cars, cities, buildings and humans. Basically called "user experience", such as in an autonomous car, all you will need to concern yourself with, in the future, is watching videos, working as if you were in your office, or simply sleeping, in complete peace of mind.

Video recognition is undoubtedly one of the major components in this new world of technical infrastructure and tools. How can we devise an autonomous vehicle in a complex human world without replicating human vision and the ability to distinguish a specific situation from its context? It has to analyze, decide and react as quickly as possible, providing an accurate, trustworthy result, and one better than human performance.

This function can be compared to the Shazam application, but extended to identify everything you hear or see in real life. This "Shazamization of my nearby environment" is a major function of the Internet of Augmented Me and there is no option other than to use edge computing solutions.

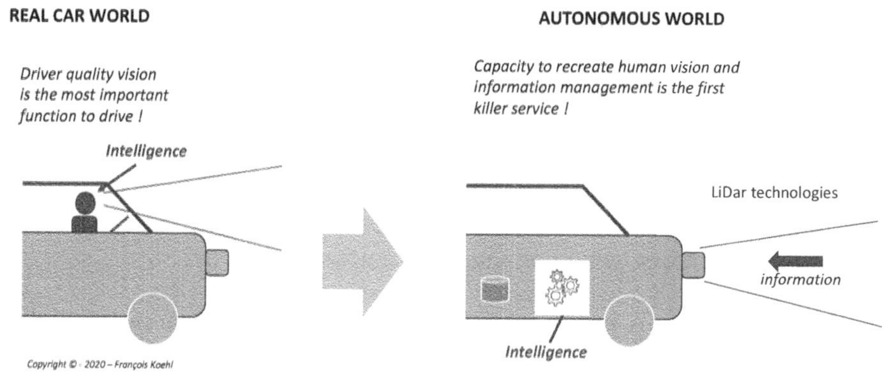

Figure 3.2. *A comparison between real world and autonomous world. For a color version of this figure, see www.iste.co.uk/duvaut/IAM.zip*

Imagine a cloudy, rainy day in the middle of nowhere, where an autonomous car will have to react as quickly as possible in unpredictable situations.

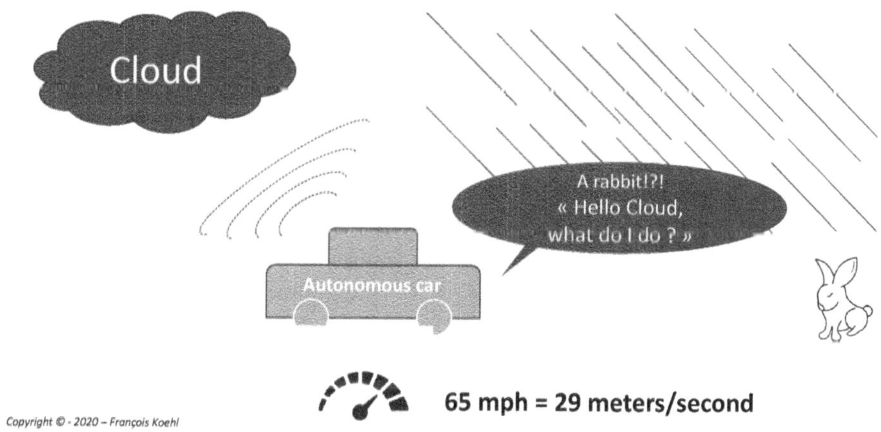

Figure 3.3. *A complex situation to predict and to treat with a 100% cloud answer, car has to integrate local intelligence to decide and act the right way in a very short time*

This is a good illustration of the importance of edge computing, because assuming that this critical level of quality of service is feasible in a fully cloud-centered world would be, to say the least, today, in 2020, dangerous or unrealistic. A car driving at 65 mph, which detects a running animal on the road, on a cloudy day in the countryside, has to react correctly, as quickly as possible, to save both the animal and the driver. How can this be achieved everywhere on earth under all circumstances with complete dependency on cloud-based solutions using non-guaranteed network latency?

Edge computing is just an unavoidable and essential piece in the digital puzzle to come.

3.1.5. *Purpose of this chapter*

This chapter aims to answer some fundamental questions for people interested in technology and entrepreneurs in the digital world. Both are concerned, or will be, with technology integration in their projects.

This chapter consolidates all of these technologies in terms of "edge computing" technology.

3.2. What is edge computing? Edge of what?

3.2.1. *Edge of what?*

The main purpose of this so-called "edge computing" is to describe an area as the technical landscape "over the Internet". It could even be defined as "what is beyond TCP/IP protocols" (presupposing TCP/IP could be omnipresent in the future) and is intelligent in itself. Figure 3.4 tries to show this generic principle.

Edge computing extends the ability of devices to work in disconnected situations or deliver services without being fully dependent on online resources.

Offline mode means operating with no connection to an API (or data stream) that would come from the Internet and would be required to deliver the service locally.

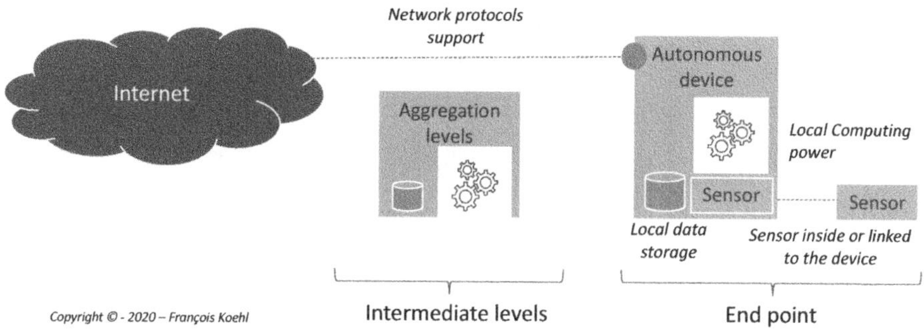

Figure 3.4. *Principle of edge computing. For a color version of this figure, see www.iste.co.uk/duvaut/IAM.zip*

Edge computing describes everything local, in opposition to what is centralized in the cloud, on the server side.

Some definitions of edge computing meet academic needs, but they are also a pragmatic viewpoint which determines the significance of edge computing in a given solution.

Few startups take edge computing's pure definition as the start point to design and build their solutions. A startup usually first spends its energy on meeting users' needs (i.e. achieving the full service even in a mobility situation), and solving technical constraints is a secondary consideration, often by iterations (i.e. making it run in every situation, even in a fully disconnected situation).

Ultimately, edge computing is an illustration of the perpetual balance between the central and local visions of computing over the years. We recall when PCs arrived to counterbalance central mainframes for performing some local and individual tasks (e.g. word processing). It is the same with edge computing. However, this is now a balance arbitrated by network bandwidth constraints (high speed connections are not available everywhere) or opportunities (5G), and local computing capacities (or "local intelligence"), these being needed in combination to respond as quickly as possible to user demands.

In summary, to give a short definition of "at the edge of Internet", we include "everything close to sensors, connected objects and users". The closer you are to "the edge", the closer you are to user experience and usage.

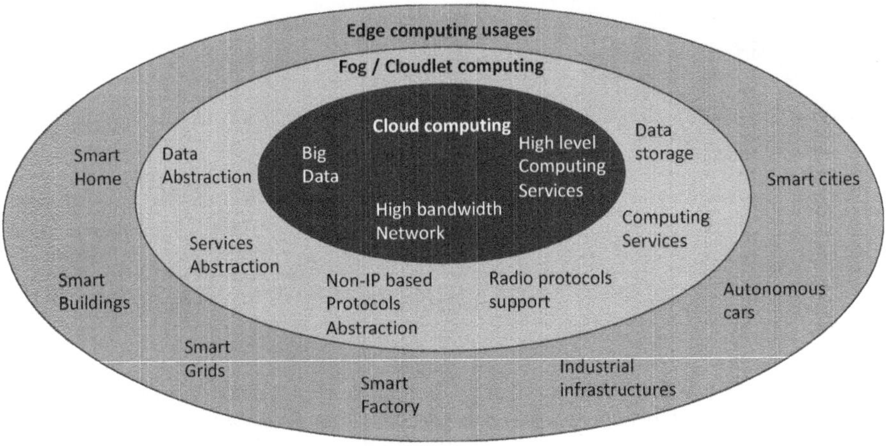

Figure 3.5. *Technologies involved in edge computing. For a color version of this figure, see www.iste.co.uk/duvaut/IAM.zip*

Figure 3.5 illustrates all the technologies that edge computing covers and integrates, e.g. COMPUTING – NETWORK – DATA.

3.2.2. *Is it a set of isolated local devices connected to the cloud?*

Edge computing is the opposite of a set of isolated local devices. It works as if all the devices, sensors and other connected objects could or will work together to contribute to a complete service. And imagine what these billions of connected local devices can generate together.

To illustrate what a set of connected local devices can generate, take Napster or Torrent as an example. These services were based on a co-operative model between all connected end-user devices. These visions of music or video sharing created a brand new notion of data management over the Internet, without cloud-centric storage, generating a disruptive view of music usage and distribution.

NOTE.– It is important to see edge computing as a set of local devices fully connected to each other, able to create value from peer-to-peer cooperation models between them.

One important point here is to consider all of these local resources as a whole, and not as a collection of isolated units only connected to the cloud.

The "whole" can equate to protection for your home, your autonomous car, or other local usages. In all of these cases, co-operation between different IoTs, sensors or local applications enables the service to be delivered to the users.

All of these local devices, sensors and other connected objects already know (or will know) how to interact together. At the edge of the Internet, technological life is not individual life but collective life. All of these units will have a specific task to perform and will create data that will enhance the potential of other local devices.

An example at the fog level of such a co-operation model is undoubtedly "mesh network" technology. This technology enables Wi-Fi endpoints to be interconnected to become Wi-Fi servers, to create a resilient and effective network.

This important peer-to-peer model has been integrated within the Linux kernel since 2008 and is used by the military to reduce dependency on Internet availability. It is a good demonstration of what edge computing can deliver.

Another example is interconnected local devices in autonomous car protocols.

At various stages of the autonomous car paradigm over the past few years, and originally, cars were imagined as fully autonomous. But to be really autonomous, cars have quickly had to consider their environment and to take it into account. The protocols designed for autonomous cars explain this reality quite well under the general name of V2X protocols, which means vehicle-to-everything. A number of versions exist:

– V2I: vehicle-to-infrastructure;

– V2P: vehicle-to-pedestrian;

– V2N: vehicle-to-network;

– V2V: vehicle-to-vehicle.

Let us illustrate this[2].

2 For color versions of the figures in the following table, see www.iste.co.uk/duvaut/internet.

V2V – vehicle-to-vehicle protocols have demonstrated the capability of autonomous cars to get information from other cars around them. Together, they create a real local network of shared information. In the event of a traffic jam for example, your autonomous car can make decisions not only on the basis of central information coming from the cloud but also using data (on acceleration, braking, etc.) from the cars around it.	**Cloud-based vision** 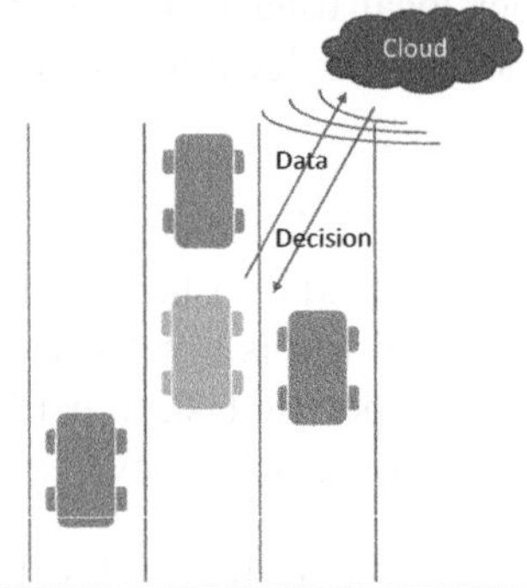
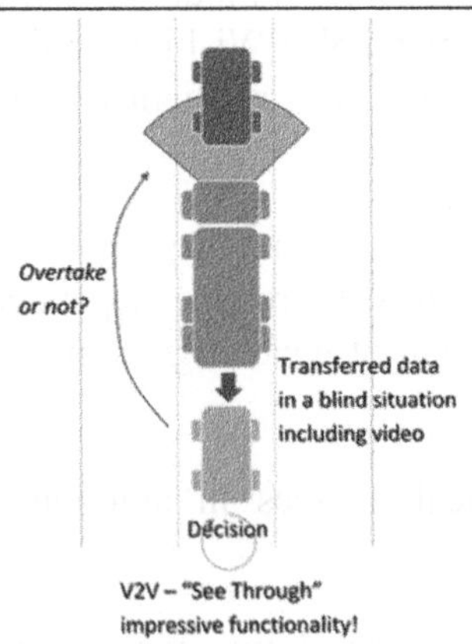 	To further illustrate the new peer-to-peer world, consider the "see through" functionality. This impressive functionality enables an autonomous car to see through the car(s) ahead in a blind situation. As the figure to the left shows, the truck ahead creates an area of invisibility, making the autonomous car unable to decide whether to overtake it. V2V protocols enable the vehicle to see through the truck immediately ahead of it and have the visibility, including video, to decide whether to overtake it or not.
V2X – vehicle-to-everything protocols were designed to integrate the ability for autonomous cars to receive data (sensor type, status, etc.) directly from components around it. Taking the same traffic jam, your car is able to discover the status of a traffic light directly from the traffic light itself.	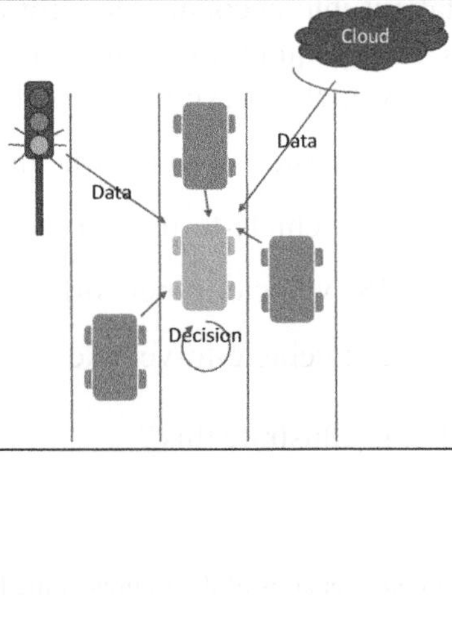

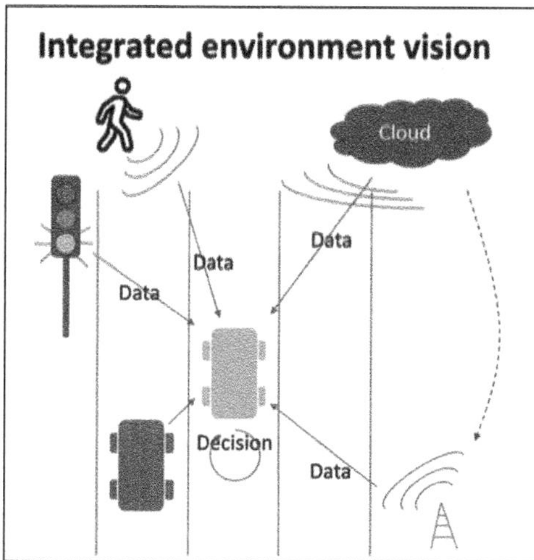	C V2X – cellular-to-vehicle-to-everything protocol communicates with a cellular network to get data from cloud-based services. It anticipates the future and the ability of cellular networks, reinforced by 5G, to serve autonomous driving processes.

This is why autonomous cars will be fully connected, and the intelligence level of cars will use and manage data coming from all connected sources through this set of protocols.

3.3. Why is edge computing so important? What are the main benefits?

It is simply this: a key driver for the Internet of Augmented Me.

Several benefits of edge computing are key to this, and the major ones are:

– it is seamless in mobility situations;

– it achieves the best reaction as quickly as possible, in real time;

– it is a crucial component in privacy protection;

– it has scalability to address volume and usage continuity.

3.3.1. *Seamless in mobility situations*

The Internet of Augmented Me has to be at the top of service delivery, either work-related or personal, for each of us, seamlessly.

There would be nothing to say about the Internet of Augmented me if it were mandatory to download data before using it, as we did in the past with video and music, for example. Before streaming arrived, to view a video or listen to music, users had to wait for files to download as MP3 or MP4. After downloading these files, you had a smooth viewing or listening experience.

Seamless means natural and continuous collaboration between all the services you are connected to at any time, from any device and from any location.

3.3.2. *Best reaction as quickly as possible, in real time*

Edge computing helps product designers by providing a continuous set of technologies from central to local, in order to decide where to process data as quickly as possible to generate the best reaction for users.

It is not only a question of balance between entirely central and entirely local computing. Edge computing, as we will see below, provides a range of technologies from central, fog, cloudlet, local and mobile edge computing.

When designing an end-user product or service, a complete architecture must be designed, from the central services, provided with APIs, to local usages using local intelligence and storages, provided by an integrated chipset.

The 2019 CES in Las Vegas clearly showed how chipsets provided by Intel, Qualcomm, NVIDIA, ZF and others integrated Artificial Intelligence, making it easier to execute high-level data processing locally, and to take decisions and react locally.

One of the best examples to illustrate the power of edge computing reactions is, in a connected autonomous car, a local computer can decide when and how to automatically park the car if the driver stops responding.

3.3.3. *Crucial component in privacy protection*

Local devices can be considered an essential component of privacy protection even though they are, at the same time, a point where privacy fails.

Because of this last mile computing, personal data are subject to theft, hijacking, unauthorized access and other attacks.

Dedicated high-level protection processing can be integrated immediately adjacent to end-users, such as face or fingerprint recognition, data encryption, double authentication factors, etc., in order to protect devices from direct hacking.

Naturally, involving local devices in end-to-end privacy protection is the best way to really protect data.

Privacy protection has to be seen both as a defensive response to last mile hacking and as a practical value requested by users. How can I trust my personal device if I am not sure of my privacy protection?

Edge computing accompanies end-users throughout their daily activities and ultimately captures and stores all the information, making it possible to know the user "better than they know themselves". This is where our digital twins start, making it impossible to overlook privacy and security concerns.

3.3.4. *Scalability to address volume and usage continuity*

We can all understand that central capacities are not unlimited and, moreover, that it is important to anticipate the ramping-up of usages to adapt those central capacities before they lack resources.

Many services have failed to anticipate their success, and been unable to serve customers because of increasing demand for resources.

In this case, edge computing, with its local processing and storage, appears to be a genuinely scalable service. Powerful edge solutions mean cloud-based servers receive less resource requests and are able to better manage their capacity plans. Because of the huge and increasing data creation from sensors and local devices, data transport becomes the bottleneck if 100% of the service is processed and delivered from cloud-based servers. Edge computing appears to be the natural solution to manage locally, to the greatest extent possible, the amount of data produced.

Likewise, a less centrally-dependent service will be less subject to the impact of a global failure. Local failure means limited local impacts that are more efficient for a distributed solution.

The benefit of disconnected usage here becomes a real advantage for end-users if the service is still delivered during cloud-based server interruptions.

3.4. Edge computing, a question of architecture and implementation models

First, edge computing is a question of architecture in a complex world of local servers, devices (from smartphones and autonomous driving computers to IoT) with a real software stack inside (local dedicated servers, or a local device playing the role of a server for others).

Second, it has to be connected to the Internet either continuously or from time to time. Edge computing focuses on the local part of a solution or service. In any event, most future devices will be connected to the Internet and will consume services through APIs.

Edge computing is a question of implementation with one main question to answer: where will you locate processing and data – centrally or locally?

To answer this question, you will have to strike a balance between:

– low latency communication at the local endpoint;

– available bandwidth, in both directions (download and upload) where large amounts of data are involved;

– geographical distribution of your service;

– the most efficient service in mobility situations;

– managing the networks available where your user is located (instead of requesting your preferred network).

In dealing with these aspects, edge architecture represents the next generation of information system architecture. It can be shown through different models, showing clearly what is "beyond the cloud" and "the nearest to user experiences".

3.4.1. *Implementation models introduction*

Figure 3.6 illustrates edge implementation models based on fog, cloudlet, mobile and local edge computing.

IFTTT is a precursive service that introduces automation through "edge computing solutions managed centrally" which will be an important part of delivered services.

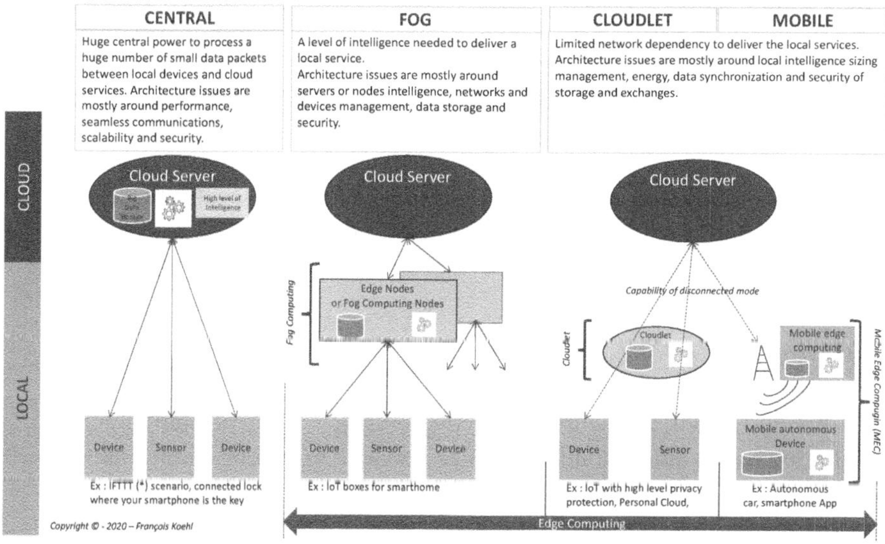

Figure 3.6. *(*) IFTTT: "If This Then That" is a 100% cloud-based application that makes it possible to control IoT from the server scenario. To switch on your connected lamp, you push the local button and the cloud connection sends a "light-on message" directly to the local lamp. The local switch does not communicate directly with the local lamp!*

3.4.2. *Fog computing model*

Designates a decentralized computing infrastructure based on fog computing nodes (FCN). These nodes can be placed between the cloud level and local devices to perform services if need be.

Fog nodes are naturally heterogeneous and include routers, switches, access points, gateways, boxes, etc. Fog nodes can be one of the local devices which provide more computing services than others.

Protocol layer: fog nodes can support several communication standards and not only IP-based devices. Fog nodes can support non-IP based communication standards to communicate between the FCN and the local devices.

Abstraction layer: one of the most important points is that fog nodes have to be as transparent as possible to the local devices. An abstraction layer exposes a set of functions as a uniform model consumed by devices or the orchestration layer.

Orchestration layer: this layer aims to use all of the available functions, from data storage to security, to serve user requests as effectively as possible.

Fog computing nodes can integrate all that is locally needed from communication, data storage, data processing, computing, security and so on, in the same way as a central server.

3.4.3. *Cloudlet model*

This designates a resource-rich computer or cluster of computers that is connected to the Internet and provides services to nearby mobile devices. It is designed as a "distributed cloud".

Cloudlet is made of real computers, or small-scale data centers, including a potential full stack of services and technologies approaching those of cloud-side servers. They are located locally and are able to deliver services with a minimum of dependencies on networks, for an optimum local user experience.

The virtual machine layer makes it possible to install and run a full stack of technologies on the cloudlet server. From this point, this is a question of the CPU, RAM, data storage, etc., needed to deliver the requested service.

For the rest, cloudlet can integrate and provide some or all of the fog computing technologies. This means these two models can sometimes be difficult to distinguish or to compare at the operational level.

3.4.4. *Mobile edge computing model*

This designates a computing resource that is able to provide everything that exists at the cloud level in a mobile device situation.

Mobile Edge Computing (MEC) has a dedicated function to deal with Radio Area Networks (RAN) which are part of 4G/5G network protocols. In this context, MEC helps to support connections and exchanges between devices or machines (Device-to-Device or Machine-to-Machine), such as an autonomous car via V2V sharing traffic density or relative speed and direction.

5G will not only mean network bandwidth or latency time. It will be a real service provider making possible new usages, some of which we cannot yet imagine. A complete chapter would be needed to explain it but it is certain that edge computing will increase on this capability to interact more rapidly and to integrate all kinds of data arriving from everywhere.

3.5. Edge computing, major technical concerns

If a business – a solution publisher or device manufacturer – has to integrate edge computing, it will encounter major technical concerns. We try to figure that out in this section.

3.5.1. *Networks*

> NETWORKS AT THE EDGE.– The greater the quality of service you want to deliver to your users, the more networks you need to manage at the edge.

At the edge, it is not so easy to cover only one type of network. Even just to implement a fix, several networks have to be tackled, and the more intelligent the device or unit is, the more network types have to be implemented.

At the very last link, in the case of a sensor, just one network protocol is integrated, often because of the energy available to run the sensor. It is at this local edge endpoint that network wars between ZigBee, zWave, Blue Ocean or others can be observed.

But when you are at the fog level, more than one protocol will be integrated – one to communicate with local devices and at least one other (such as Wi-Fi or 5G) to communicate with cloud-based servers.

One of the more complex situations is C-V2X, where a high number of protocols have to be integrated to serve a large number of sensors and create resilient usage in mobility situations.

The number of possible network protocols can be so complex that a Network Function Virtualization (NFV) is now emerging, bringing a real vision of "Network-as-a-Service" to make network integration easier.

3.5.2. *Security*

SECURITY AT THE EDGE.– Edge computing is both a new vulnerability endpoint (the poison) and a new potential active agent (the antidote) at the frontier of the Internet, helping to enhance the overall security level.

It is clearly a local vulnerability endpoint because of the number of distributed and, most of the time, autonomous devices. The main question for edge architects is: what is the best way to protect a large number of devices or connected objects?

Some vulnerability points are as follows:

– autonomous cars can be managed remotely if hacked;

– hacking is less easy to detect because of local unit management;

– using the peer-to-peer co-operation model (see above) means a hacked device or IoT unit can propagate malware rapidly, in the same way as connected PCs in a local network when there is no firewall;

– stored data are limited to a local context (personal, family, etc.) but can easily be used if hacking occurs.

At the same time, edge computing can help secure the Internet frontier, as you can see in the following cases:

– local data storage means personal data can stay local, avoiding a pointless and potentially unsecured concentration of personal data at some central point;

– local computing capacity means edge computing is able to anonymize or to encrypt data before sending it to a cloud service;

– local computing capacity combined with network protocol management means edge computing can help to create local security agents. These agents can be integrated with edge solutions, routers, networks and even devices or sensors, and can collaborate together either at a local level or to send/receive information from cloud-based servers.

Publishers and technology firms such as Microsoft, Google, Amazon and others are preparing a security framework for the IoT that will be a new, decentralized security management system.

3.5.3. *Energy management*

ENERGY MANAGEMENT AT THE EDGE.– Any talk of autonomous and sustainable local devices implies managing and optimizing energy consumption in delivering network management, data storage and computing processing power.

Local devices and the IoT need specific energy management. Edge computing, due to the local services of location management, network flows, data storage and computing, needs energy management. The more local and intelligent the local device is, the more difficult the question of energy becomes.

Local energy management is well known by smartphone users. Everyone knows that using a localization app such as Plan, Google Map, TomTom or Waze consumes so much power that you have to be close to a power supply or you could be lost in the middle of nowhere.

Where autonomous sensors run off a built-in battery is a real concern for IoT management. Thinking about the large number of IoT in a house within a few years, battery replacement or recharging is a major issue to be solved.

Edge computing architects have a difficult job measuring all consumption of the chosen components to strike the best possible balance between service delivery and autonomy. Let us take the last-end network protocols as an example. At this point, IoT and sensors use protocols such as ZigBee or zWave and do not use commonly-used protocols such as Bluetooth or Wi-Fi, because the energy consumption to maintain pairing and exchange data would be too high and would create constraints on making the solution sustainable.

However, some edge computing solutions can take advantage of their context. For example, in an autonomous car, an energy source will be easy to find and to share with the car engine. The energy consumption of edge computing solutions will be but a small fraction for any autonomous car. To take another example, the Hue lamp from Philips is directly plugged into an existing power supply, making it possible to support protocols such as Wi-Fi to connect to a box and even making it possible for it to become a Wi-Fi repeater.

3.5.4. *Combining technologies to react at the right speed*

COMBINING TECHNOLOGIES.– Edge computing demands that different technologies are combined to get the required result, often using a trial-and-error approach, and the result is generally a more complex solution than the one first imagined as the end-to-end solution.

As we saw in the implementation models, edge computing solutions need a real architecture view to design and adapt the right solution. Combining technologies for data storage, data analysis, network management and artificial intelligence to generate end-user solutions is quite a complex puzzle to design, integrate, build, test and maintain.

There is no real model to determine precisely what the global end-to-end response time is likely to be. You design your solution as best you can, but the real speed will be measurable only at the end of the architecture proof of concept. This means that in many cases, you will design, test and then adapt to finalize your solution design under an agile project management process.

That is where edge computing models will be important because you will have to balance between local and central across edge computing, data storage and the network layers overall. To add to the difficulty, you will have to define precisely what will be built as software and what will be used at the hardware level.

As we saw above, chipsets have now integrated a high level of artificial intelligence processing, creating a real local framework and avoiding code development or deployment.

All these key technological dimensions are summarized as follows:

– choice between local, fog/cloudlet or central;

– data storage/computing at each level;

– network protocol management;

– software or hardware management;

– local energy management.

Finally, the real key performance indicator is how the user feels about their experience. This subjective speed will be one of your nightmares because some users could stop using the service if they consider it "too slow". This subjective speed is not only the result of your product's quality but also the usage context, and, mainly, network quality and latency if your service depends on central data being downloaded. This is why edge computing will be a good solution to create disconnected remote usage without networks. However, it will also make your solution more complex.

3.5.5. *Artificial intelligence integration*

> ARTIFICIAL INTELLIGENCE AT THE EDGE.– Because of its local usage, edge computing has to be ready to handle many situations, both to analyze them and to react to them. That is why many edge computing solutions will, in the future, integrate artificial intelligence as close to usage as possible (known as "Edge AI" or "AI at the edge").

When designing a central solution, anticipating the usage conditions and monitoring them to make the necessary adjustments is quite straightforward.

Central solutions run in datacenters where the best levels of support, energy consumption and computing capacities are found. You can even easily imagine enhancements with new features, new resources, etc. Implement once, and every user will benefit.

At the central point, if a technical computing process fails, you can fix it or, sometimes, run it manually to correct it or to avoid visible impacts for end-users.

However, at local endpoints, the same does not apply. Your solution will have to cope with many different situations to react independently. This is where artificial intelligence becomes important. Taking all the available data describing the context of usage into account, artificial intelligence will help to make the right response to the usage, providing a real high-level user experience.

In the early days of your solution, this might not appear to be a major concern. But the more users use your service, the more data you will need to manage, and the more artificial intelligence will appear as a killer service to tackle complexity. Anticipating this point could be an accelerator for you if success comes knocking!

3.5.6. *Software development practices*

SOFTWARE DEVELOPMENT PRACTICES AT THE EDGE.– The number of technologies to be combined to create an end-to-end architecture means that software development practices will be many and varied with particular constraints to solve at each level to ensure overall quality of services.

As we have seen, edge computing is a mix of technologies. Taking that into account, it means that an edge computing solution must deal with the different pieces of the puzzle, and then adapt development practices to this new and complex environment.

To succeed, it will be important to master a complex stack of technologies working together. To do so, you need to be aware of different technologies, and adapt your team as well. It is a time of talent renewal: CTO, developers (from front, back or full stack), statisticians, information system architects and so on. And each of them has to be able to develop software in their own technical area and co-operate with others too.

This is not so different from software for industrial environments or embedded software. But the main difference is that the technical endpoint co-operates, through different network protocols, with different levels of information from cloud to local through the fog level. This must be synchronized and secured at each stage of the architecture.

The edge computing development team will have to deal with:

– the number of technologies to integrate, with new and not yet standardized options appearing almost every day;

– the imminent new versions of IoT secured frameworks in their infancy, without much support or feedback as regards deployment at this point;

– Solution Life Cycle Management with complex processes to manage at the local level such as versioning, maintenance, monitoring, bug discovery/reporting, bug fixing, etc. All the conventional aspects of software development must be anticipated and automated, and must take into account the distance between the local devices or sensors and the central management point;

– the blend between software and hardware will be specifically managed first, to build the solution, combining the strong points of software and hardware. But at RUN time, it will be very specific, primarily because of cybersecurity. When a security flaw is discovered in a software module, it can be quite inexpensive to fix it and deploy the new version. But what if the bug is discovered at the hardware level? Can you fix it with the software? Can you avoid using the functionality in question? Is your endpoint solution corrupted?

As observed, development practices will evolve to match the complex and dynamic situation prevailing at the edge.

3.6. What about usages?

FUTURE USAGES.– Because of its powerful local capacities, edge computing allows all devices to become a service delivery unit. A new paradigm appears: "Everything as a Service" from cloud-based servers to local autonomous devices.

In this section, we do not claim to explain all the use cases based on implementing edge computing. We choose to illustrate some characteristics in order to explain how to look at these business cases from an edge computing point of view.

In all these business cases, edge computing helps to distinguish the cloud-server side (central) areas from the areas of local experience (local) at the endpoint. Edge computing contributes to end-to-end solutions, and transforms the "Internet of Augmented Me" into a responsive, personalized and omnipresent service.

The term "areas of experience" helps to designate UX solutions and to identify the value added by local edge computing technologies. Edge computing's role in the user experience, in this case, is major.

An example is the connected city of New York where water flows are fully managed from all the main water distribution points to central management. At the endpoint, New York residents can monitor their own water consumption from a local app. You can easily imagine the complex networks of sensors and the amount of computing power needed to manage all these data.

The main question is: from the business viewpoint, how can an edge computing stage be included in your projects, in your business model? And how will you manage technology from the endpoint view within a collaborative model?

3.6.1. *Smartphones*

Smartphones and the like, including tablets, are the new paradigm of what was the personal computer (PC). At the time, the PC was an item of technical equipment (typically in the office or home), found where smartphones or tablets were later to become real access points for the "digital twin", supporting and serving everyone at every stage of their lives.

Computing	It is no secret that each new smartphone version includes a new generation of CPU (central processor of your smartphone), showing how important it is for a smartphone to be able to process complex operations. For example, the recent Apple A12 bionic CPU in the iPad Pro is comparable to Intel i7 8th Generation. Today, you have more computing power in your pocket than was needed in 1969 to put men on the moon!
Data	The difference between smartphones, for the same model, lies in the quantity of data you can store. It is less a question of data storage than of "seamless mobility experience". The volume of data stored on a smartphone explains exactly what edge computing is: "the more data you can store, the less dependent you are on a permanent Internet connection while moving around". For example, the data capacity to store a complete TV Series on a smartphone such as *Game of Thrones* in order to watch it during a flight helps explain the role of data in disconnected mobility situations.
Networks	Ultimately, for all versions of smartphones, network support is a key point. When 4G was introduced, we could easily see the difference from 3G. 4G has made a new range of applications possible, such as Netflix or Molotov, able to stream series or movies in a mobility situation. In the same vein, 5G will be a key driver of choices for new smartphones in the next two years. Because 5G will create new types of application, for example, augmented reality mixing your vision of reality and information coming from a cloud-based server in a real-time experience: in a crowd, you will be able to recognize a friend and get all their recent posts on Facebook, Instagram and so on.

3.6.2. *Internet of Things (IoT)*

The Internet of Things is a complex world which encompasses many different situations. Let us consider it as a whole: the Internet of Things is the world of sensors and connected objects. These sensors supplement solutions where users find a real experience and high-value services. The Internet of Things is truly a new paradigm where it is no longer the user who connects to the Internet, but in fact an object operating autonomously.

The IoT is participating in completing the solution (smart home, smart city, etc.). That is the reason why the term is now so global, embracing all the solutions. To illustrate the IoT, we will consider the last mile object independently from the overall solution of which it forms part.

Let us call it the "last mile object", or LMO.

Computing	Computing power is less significant for the LMO than in other places of the technical chain. Why? Because the LMO is a world of connected and autonomous objects identified as units. This local object has to be autonomous for energy management, as most of the time it is disconnected from a power source and some distance away. But the more intelligent the connected object is, the more power it needs. Power is important for the billions of connected objects planned for the future. The LMO will be more a matter of data and networks than computing. That said, taken in the round, computing will become a concern at fog and cloudlet level; see below on smart buildings, smart cities and smart homes.
Data	Data is a real concern for LMOs, but more due to the unit creating data than as a storage necessity. Most connected objects, based on sensors, create a data flow, and their primary role is not to store data but rather to transmit data to a higher level in the system as soon as possible. Here, the data sent by LMOs are often simple and sent in small packets. This data's value can come from a single packet for an event (e.g. a fire alarm) or from a large number of packets (e.g. several LMOs in a city to map the traffic).
Networks	Network protocols are the real concern. In a cable or wireless connection, the chosen protocol has to be resilient for the LMO. If the network is not fully operational, then the connected object could become isolated because of lack of storage, as we saw before. In such a situation, because it has not been designed to store data, it could lose one or more datasets. Consider a movement detector sensor built into an intrusion solution, with no operational network. It will not be possible to send event data to the central box and trigger the alarm. The chosen protocol is also a power issue, because protocols are not all equal in terms of power consumption from the pairing process to the transfer process.

3.6.3. *Mobility: autonomous cars and vehicles*

We have already seen the extent to which autonomous cars represent edge computing. It requires local intelligence, data storage and network support. Autonomous cars will be one of the most complex objects in the galaxy of edge devices. From its ability to resolve complex problems in dramatic situations to its ability to take into account all of the available data coming from all of the connected objects or data sources within its environment, an autonomous car represents an amazing amount of everything: computing, data and network capacities!

<table>
<tr><td>Computing</td><td>As we saw previously, autonomous cars will be a real local and mobile data center, including a complex set of servers and a high level of local intelligence. NVIDIA or Qualcomm and other chip factories such as ZF are endeavoring to integrate as many computing facilities inside cars as possible, including artificial intelligence services. Tesla provides an impressive example. Tesla built a solution named "Full Self-Driving Computer" considered to be one of the most powerful processors in the world. It is quite difficult to imagine what 144 teraflops are inside a car. By way of explanation, if you conduct one operation per second, then to process the same number of operation that 144 teraflops perform in 1 second would take you 4.5 million years! But this high-level of computing has a price; it consumes 72 watts (vs. 57 watts for the previous Tesla version with NVIDIA), creating a trade-off between self-driving and the real autonomy of the car.</td></tr>
<tr><td>Data</td><td>As you can imagine, from data creation (Lidar, imaging, videos, etc.) to data flows coming from cloud-based services or other cars or environment data (weather, information on traffic conditions, the status of cars around it, etc.), an autonomous car has to handle an amazing flow of data, whether to store it, to analyze it or to exchange it in both directions. In terms of volume, just imagine a car as an integrator of hundreds of sensors producing many terabytes of data per day. To take a clear example, volume is between 20 and 100 Mbit/s for Lidar (1–5 per vehicle) and 500–3,500 Mbit/s for a video (6–10 per vehicle), according to Lucid Motors. This will be a key challenge for car manufacturers to resolve: processing data as effectively as possible and identifying – from that volume of data – the most valuable data and how best to manage them.</td></tr>
<tr><td>Networks</td><td>Autonomous cars will certainly be one of the most connected objects in digital life. Because of their mobility, network protocol support is a twin challenge. Firstly, it is important to be as resilient as possible, using all the available networks. This is why autonomous cars will support as many network protocols as possible, namely 3G, LTE, 5G and so on. Secondly, network protocols give location information. The available networks help locate the vehicle more precisely, supplementing satellite information (GPS, Galileo, etc.).

The old generation of network technologies made it possible to locate a vehicle within a radius of five meters. Combining several technologies, new systems can locate a vehicle within one meter. This is crucial to taking the right decisions.</td></tr>
</table>

3.6.4. *Smart building, Smart city, Smart home means Smart everything*

The "smart everything" family is a way of organizing technologies creating user experiences in all areas of life. Edge computing is an essential component in the user experience, fully seamless through all the "smart everything technologies" and the collaborative models which can be created. Figure 3.7 illustrates this point.

It is important to understand that all of these solutions are in the infancy of their future development. Everything is built separately but at the same time. The main standards at each level are not defined yet. Now is still the time of "early adopters", and many of the startups arriving in these marketplaces still rely on the well-known "first-mover advantage" effect.

Business profit viewpoint: these business cases are often initially built independently. However, to ensure business development and provide a real user experience for their customers, they need to co-operate with other services. This creates technical solutions combining different technologies, as previously explained in the IoT section. This could create an excessively complex business model where customers do not want to pay for the stack of services needed but do ultimately want to get an affordable, complete service.

Non-profit organization angle: in contrast to the business profit motive, all of these solutions can co-operate to share information under a non-profit business model. For example, a smart garden can measure the quantity of rain and the soil moisture level. This information can be aggregated at the smart city level to optimize the watering of municipal parks and gardens. Real citizen contributions will be crucial in the future to avoid a multiplicity of the same sensors in a small geographical area. Such solutions make it possible to build an "Internet of Plants", which could help to manage biodiversity.

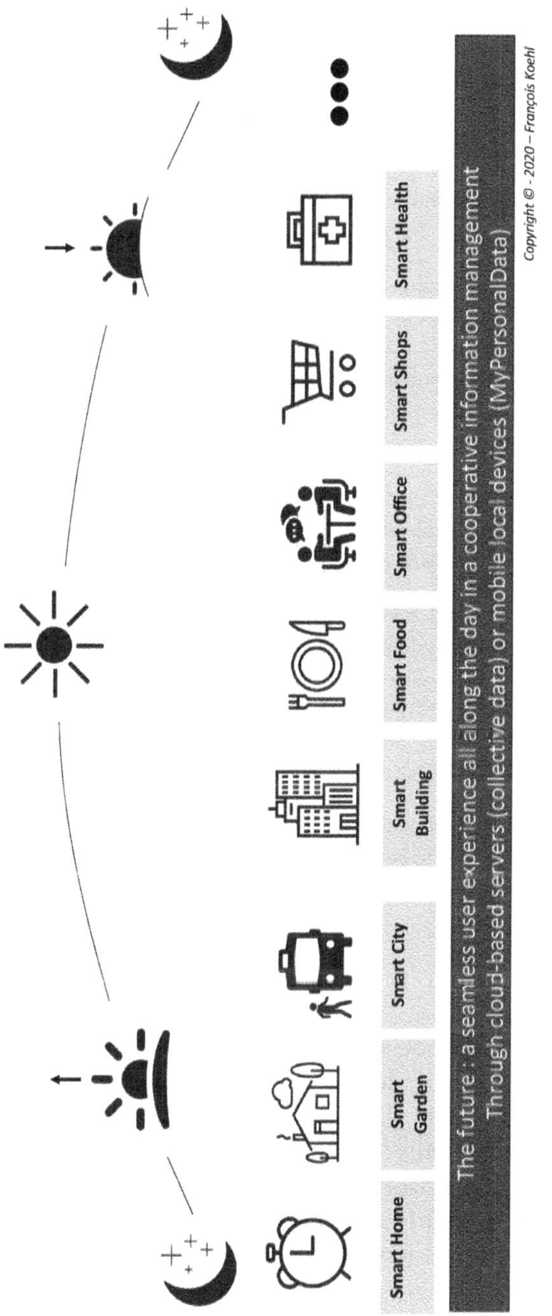

Figure 3.7. *Smart everything*

Computing	A wide variety of technologies: – existence of local servers or boxes in some location. These local servers or boxes have some key functionalities such as network protocol management (more than one to integrate IoT), sensor management, security management, data storage and analysis; – data flow analysis to detect specific events and respond at the right level; – autonomous operation with no network (disconnected mode) or without power (autonomous mode).
Data	It is important that data are fully captured (from sensors), and stored and supplemented (from APIs) with data from cloud-based solution. In these "smart everything" solutions, you will find all of the implementation models explained above. Data will be managed at every level from central to local and at the fog level. To build such solutions, it is important to think of data as a distributed system consuming external APIs to complete the service.
Networks	Smart everything is a world of heterogeneous networks. It is difficult for a solution in this field to choose a specific network protocol without risking obsolescence or interoperability problems. It is important to be able to support different protocols at fog and local levels, to be as open and scalable as possible.

3.6.5. *Industry*

Industry is generally not naturally associated with edge computing. However, for a number of decades now, industry has been building a complete information system (IIS – Industrial Information System) which is locally managed and centrally controlled. Inside factories, for many years, sensors, SCADA (Supervisory Control And Data Acquisition), DCS (Distributed Control System) and other supervisors have pushed data from one endpoint (the industrial device) to a central point, in order to optimize processing, maintenance, failure detection, prevention, etc. For some time, this Industrial Information System has been linked to suppliers via the Internet and through a highly secure link, and to the Enterprise Information System via the corporate network. This means IIS works in an edge computing environment.

We highly recommend reading Chapter 5 on "augmented industry" to understand how computing, data and networks contribute to this new age of manufacturing.

3.7. Takeaways

3.7.1. *Main definitions*

EDGE COMPUTING IN A FEW WORDS.– Edge computing represents everything performed as close as possible to data creation and usages. While it complements them, it is the opposite of all those tasks performed centrally, at the cloud level.

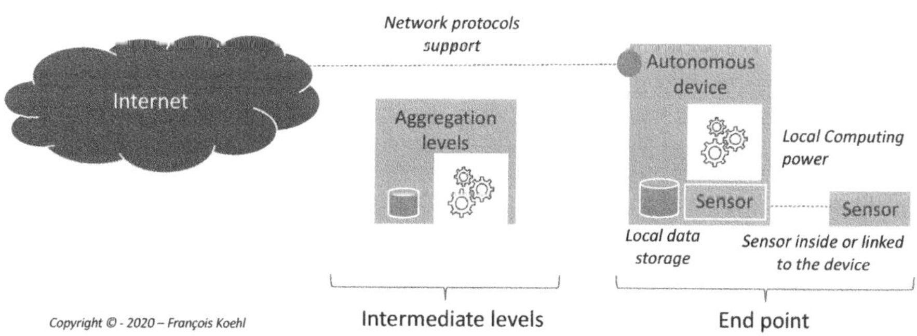

Figure 3.8. *What edge computing represents at the endpoint. For a color version of this figure, see www.iste.co.uk/duvaut/IAM.zip*

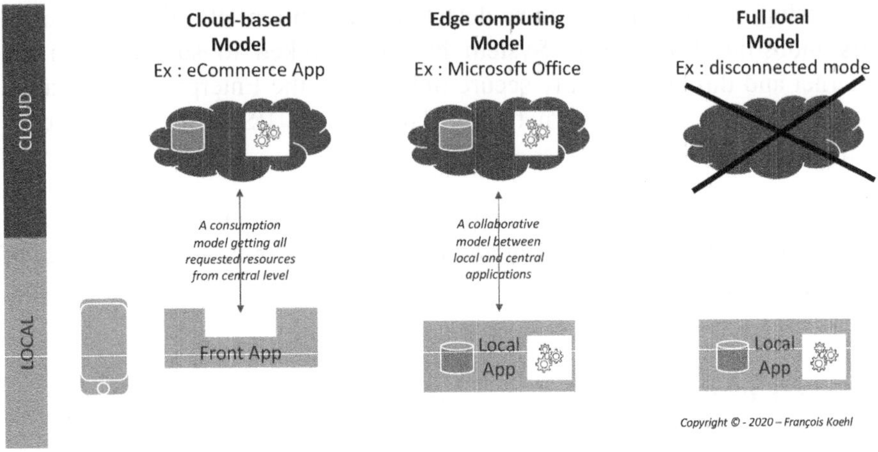

Figure 3.9. *Different cloud-server and local usage implementation scenarios. For a color version of this figure, see www.iste.co.uk/duvaut/IAM.zip*

Edge computing is the natural new age in the continuous evolution of information systems architecture.

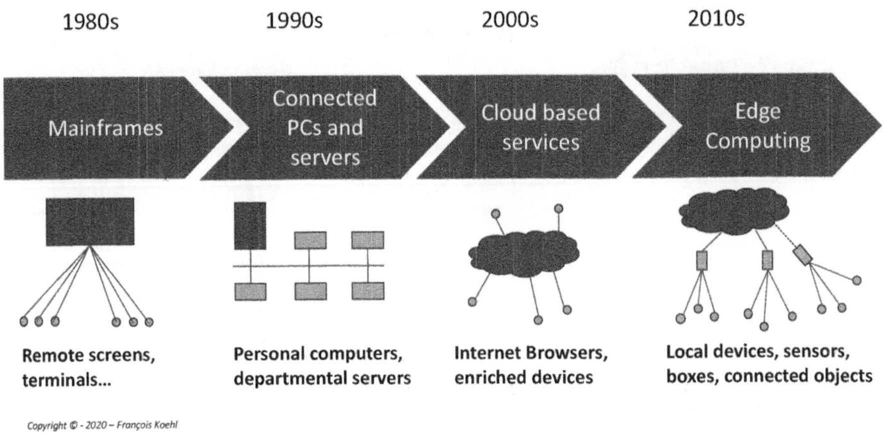

Figure 3.10. *The continuous evolution of information systems architecture. For a color version of this figure, see www.iste.co.uk/duvaut/IAM.zip*

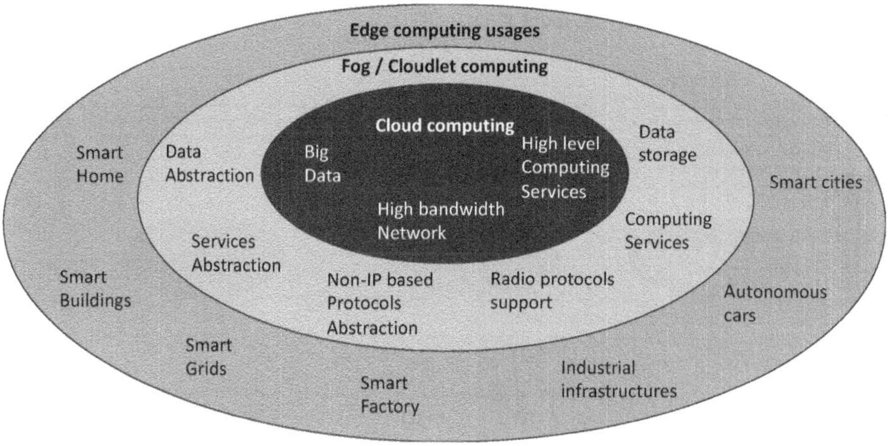

Figure 3.11. *The different levels of components in the edge computing landscape. For a color version of this figure, see www.iste.co.uk/duvaut/IAM.zip*

The main benefits of edge computing

Seamless in mobility situations	The ability to use your local service in any situation and environment due to local capacity to process data and react.
Best reaction as fast as possible, real time	The ability to react in near real time due to local capacities without network dependency.
Masterpiece of privacy protection	The ability to bring local protection or correction services to avoid cyberattacks, protect data and limit attack impacts.
Scalability to address volume and usage continuity	Local devices, able to manage data locally without consuming too much server-side resources, are more easy to multiply, creating a scalable solution

Edge computing architecture models

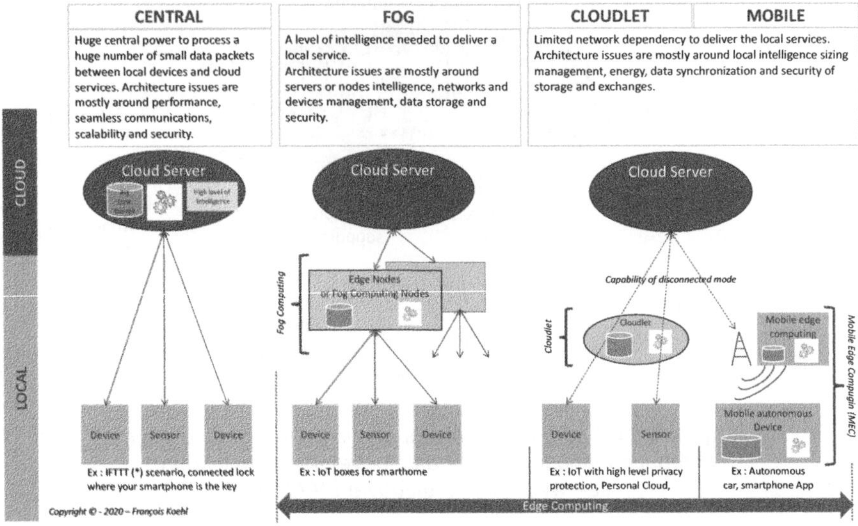

Figure 3.12. *For a color version of this figure, see www.iste.co.uk/duvaut/IAM.zip*

Fog computing	Design a decentralized computing infrastructure based on fog computing nodes (FCN). These nodes can be at any need place between cloud level and local device to perform services.
Cloudlet	Design a resource-rich computer or cluster of computers that is connected to the Internet and provides services to nearby mobile devices.
Mobile edge computing	Design a computing resource making it able to provide everything existing at the cloud level to a mobile device situation.

Major technical concerns

Network	The higher quality of services you want to deliver to your users, the more networks you need to manage at the edge.
Security	Edge computing is both a new vulnerability endpoint (the poison) and a new potential active agent (the antidote) at the frontier of Internet participating to enriching the global security level.
Energy management	Speaking of autonomous and sustainable local devices implies managing energy consumption to make possible, with minimum constraints, network management, data storage and computing power.

Mixing technologies	Edge computing imposes mixing different technologies to get the required result in a try and learn approach, and the result is generally a more complex solution you first imagine if you look at the end-to-end solution.
Artificial intelligence	Because of its local usage, edge computing has to be prepared for a lot of situations, to analyze them and to react as well. That is why any edge computing solution will integrate artificial intelligence as close to usage as possible ("Edge AI" or "AI for edge").
Software development	Because of the number of technologies to mix in order to create an end-to-end architecture, software development practices will be multiple and heterogeneous with particular constraints to solve at each level to ensure global quality of services.

CONCLUSION.– Edge computing will help in any new solution if you want to deliver a high-quality user experience. It will be applied in many domains, your smartphone being foremost among them. The Internet of Things will bring many data flows that will enhance local and cloud-based solutions. The autonomous car will be a major example of edge computing with one of the most complex locally-managed and fully-mobile systems. All of this will work towards "smart everything", which will cover all of the situations you encounter on a daily basis.

That is why edge computing is an essential piece in the puzzle making the "Internet of Augmented Me" possible.

3.7.2. *Edge computing key performance indicators (KPIs)*

A KPI, Key Performance Indicator, aims to measure the capability of the edge computing technology to meet their goals. Here are some reflections about it.

KPI – Network consumption: avoiding cloud obesity and network saturation

Network consumption is one of the main KPIs for the edge computing strategy. It gives the autonomy to store, process and use local data without making cloud requests as a matter of course. Edge computing will preserve Internet data flow for important usages, distinguishing a real necessity to consume data over the Internet from nice-to-have usage.

KPI – Best user experience level in real time

Every user is able to understand edge computing when their mobile device reacts in real time in any location and any kind of situation, with no latency, taking all the constraints or opportunities of the user's context into account with a personalized response, and avoiding such messages as "No Internet support. Please connect your mobile device to get the full service." The best UX will be in real time, seamless and continuously adapted to the user's context.

KPI – Resilience

One of the non-visible KPIs of edge computing will be service resilience. Why non-visible? Because when a service works perfectly, every user finds it "normal". The user only takes notice when the service is down because of a technical failure or network unavailability. This is even more true when major failures occur.

KPI – Openness to multiple technologies and data flows

In an ideal world of edge computing, proprietary solutions would work together. It is easy to grasp, but not so easy to make happen without common standards. An edge solution will, in the future, be open to integration or interaction with several technologies and data flow. The more open it is, the more external services it will be able to orchestrate, and the more valuable the services it will deliver.

KPI – Sustainability

One of the main risks of Edge Computing is creating new sensors, new devices and, as we can already see in the smartphone market, generating a strong obsolescence rate. This risk is real considering the 75 bn connected object to come in 2025! One of the KPIs of edge computing will be in terms of energy consumption and obsolescence management. For energy consumption, edge computing could integrate new solutions for energy collection and create a new category of "battery free" objects. We can already see examples on the market with solar power support or energy generation coming from physical movement. For the obsolescence rate, openness to multiple technologies will be crucial to avoid a multiplicity of vendors with proprietary solutions. We all know that it will be a hard battle against the goal of market increase.

3.7.3. *Edge computing – main figures*

Edge computing is a world of "large numbers", whether in terms of data storage, the number of data flows involved, the number of objects (sensors, smartphones, etc.), market size and growth, and so on. Any figures around edge computing are already huge and will increase rapidly. Some examples are presented to illustrate this point.

Data management

Considering a survey from Vodafone, 75% of data will be on the edge in 2020 via the Enterprise Network. Referring to this Enterprise Network, Vodafone explains that the market will evolve from a centralized cloud to a distributed cloud, and it expects 75% of enterprise generated data to be processed outside of a centralized data center by 2020.

Considering several forecasts, between 5 to 20 TB will be generated per day for 1 autonomous car. As of 2020, many cars are still in a test mode where data capture and analysis are crucial to developing new AI models. These models will certainly be refined in the future. However, the car industry anticipates several terabytes per day per car and that is why autonomous car is a data-driven market.

Market size: as for the market size, the trends indicate around $3-5 bn by 2020:

– according to Grand View Research, in its Edge Computing Market Size & Forecast Report, 2016–2025, the size of the global edge computing market is estimated to be worth $3.24 bn by 2025;

– for CB Insights, Market Sizing tool, the global edge computing market is estimated to reach $6.72 bn by 2022.

As you can see, it is possible to find different opinions on the future of edge computing. Even though the forecast numbers might be different or simulated, we can say that all such surveys consider this market to be growing rapidly and generating increasing volumes of everything (data, networks, billions of dollars, etc.).

But this growth will take the form of a tangled jungle of technological solutions and a mixture of quantities of usages which will make it complex to analyze it as a whole.

3.8. Glossary

APIs: connection point and data consumption model for getting data from a server or returning data to a server.

Architecture: the way to design a complete solution based on servers, local devices, information and data management, etc.

Cloud-based: everything performed server-side, by pooling resources between several users.

Cloudlet: designates a resource-rich computer or cluster of computers that is connected to the Internet and provides services to nearby mobile devices.

Digital twin: all the information you capture or store on the same data subject, in order to be recognized and delivered the best personalized service. This information creates a digital persona which is the mirror image of the human user.

Edge computing: everything performed as close as possible to data's creation or use. While it complements them, it is the opposite of all those tasks performed centrally, at the cloud level.

Edge data: represents data stored within edge computing components, in fog computing, cloudlets or locally.

Fog: designates a decentralized computing infrastructure based on fog computing nodes (FCN). These nodes can be placed between the cloud level and local devices to perform services if need be.

Implementation: the practical form taken by a solution during the build phase and during the run phase (operation phase), bringing the service to the end-user.

Mobile edge computing: designates a computing resource that is able to provide everything existing at the cloud level to a mobile device situation.

Network bandwidth: the capacity or data transfer rate of a network which represents the capacity of a network communication link to transfer data from one point to another (measure per second).

Network latency: the delay that happens in data communication over a network. The longer this delay is, the worse the network connection is. It tends to create a poor user experience.

Open data: dataset coming from an open source, giving information about context (political, economic, geographical, etc.) and thereby enhancing a service.

Scalability: the capability of a system to be used in a range of situations and to deliver the same service when the number of users or volume of resources managed dramatically increases.

4

Blockchain for New Flows of Value and the Empowerment of Me

"The difficulty lies, not in the new ideas,
but in escaping from the old ones."
John M. Keynes

4.1. Introduction

4.1.1. *Bitcoin, the mother of blockchains*

Welcome to a new world of blockchain and blockchains.

Understanding how blockchains operate can be a bit tricky. Yet, it's essential to grasp their capabilities and the philosophical, cultural and ideological principles that underpin those networks.

And the best way to start the blockchain journey is to recount the history and the values behind the first live (and 10+ years old) blockchain, the Bitcoin blockchain. As stated in the Bitcoin founding whitepaper, the goal of this new system is to create a digital cash emancipated from financial institutions [NAK 18]:

> A purely peer-to-peer version of electronic cash would allow online payments to be sent directly from one party to another without going through a financial institution. Digital signatures provide part of the solution, but the main benefits are lost if a trusted third party is still required to

> prevent double spending. We propose a solution to the double-spending problem using a peer-to-peer network. The network timestamps transactions by hashing them into an ongoing chain of hash-based proof-of-work, forming a record that cannot be changed without redoing the proof-of-work.

By solving the double spending problem without relying on a financial institution or any other kind of third party of trust, Bitcoin is cutting out the middleman, thus empowering citizens and more generally Internet users to carry out transactions without requiring any permission from any organization. Trust is shifting from central institutions to a decentralized network of computers called "miners".

4.1.2. *The consequences of Bitcoin or the Internet of Value*

Bitcoin can be thought of as a layer of technology built on top of the Internet and its innovation has several implications: for the first time in digital history, digital scarcity is possible and digital value can be shared over an open public network like the Internet. The Internet is no longer only meant for the sharing of information; it's also now about sharing value digitally.

The "Internet of Information", as it is experienced by sharing images or emails, has now been upgraded to an "Internet of Value": sharing digitally unique objects and the value they represent.

The empowerment offered by this Internet of Value comes at many levels as shown in Table 4.1.

Power architecture	Power is shifting from platforms (trusted third parties) to the edges (end users).
Identity	Any user can generate their own identity without requiring permission from any institution.
Censorship resistance	No entity has the ability to block a transaction permanently or to seize cryptocurrency balances.
Transaction speed	By cutting out the middlemen, transaction speed can be improved.

Transaction cost	By cutting out the middlemen, transaction cost is lowered. Micro-transactions become economically viable.
Risk	By cutting out the middlemen, the counterparty risk is being removed.

Table 4.1. *Empowerment levels enabled by Bitcoin and other blockchain decentralized network*

A Bitcoin-like network subsequently opens up many new (empowering) possibilities such as:

– *Banking the Unbanked*: a Bitcoin-like network can represent an alternative financial system for the 2 billion people without government aid and unable to access the banking system. Early adopters of Bitcoin (and other cryptocurrencies) have already started to benefit from this open network by generating their own identity.

– *Blockchain of Things*: programmable money like Bitcoin is opening up new scenarios of financial transactions between connected devices. For instance, we can envision autonomous vehicles also being financially autonomous due to their ability to buy or sell services.

4.1.3. *From Bitcoin to blockchain(s)*

Bitcoin has introduced a new trustless way to operate distributed ledgers without requiring a middleman. The Bitcoin ledger is a financial ledger keeping the balance and history of all Bitcoin units of account transfers. Other blockchain networks may however handle other kinds of ledger data.

Generally speaking, a blockchain is a network that allows untrusted parties to reach consensus on a shared digital history without a middleman or a third party of trust.

It is important to realize that blockchain networks are not just a technology. They are a combination of cryptography, community, social consensus, economics and more.

Let's unpack the typical layers of most blockchain networks, as shown in Table 4.2.

Cryptocurrency	Cryptocurrency (sometimes called tokens) is the most visible and probably the most exciting part of blockchains. Cryptocurrency is the economic incentive for miners to contribute hardware capacity of the blockchain network, thereby creating the untamperability of the network.
Consensus algorithm	The consensus algorithm is the mechanism or method that allows the network participants (called minors or validators and sometimes bakers) to agree on a single truth of the network's state and transactions. Bitcoin introduced the Proof-of-Work (PoW) algorithm but several other algorithms are currently under development to achieve the same level of security while requiring less computing power and energy consumption.
Peer-to-peer network	A blockchain network implies a (decentralized) physical computing infrastructure provided by miners (sometimes also called validators or bakers).
Blockchain client	Each node of the decentralized computing infrastructure is running a software binding all the computers using the same consensus process.
Open source core developers	Blockchain software (client and tools) are generally open source. This enables a better security and is a key requirement for collaborative innovation and for building a strong ecosystem.
Development platform	Blockchain, allows developers to build applications (called *Dapps* as *Decentralized Apps*) on top of the infrastructure. Permission-less innovation is unleashed!

Table 4.2. *Blockchain, unpacked typical layers*

4.2. The six characteristics of blockchains

Six letters can define the main blockchain characteristics: I.M.P.A.C.T.

It is important to note that the person who first formalized this acronym was Xavier Dalloz, co-author of this book.

4.2.1. *Identification*

Blockchains use cryptographic authentication and signatures to authorize any operation; the blockchain keys are the blockchain identities. For instance, the e-residency program allowing digital citizenship in Estonia was built in partnership with Bitnation, a blockchain-based identity solution.

4.2.2. *Money*

Crypto-assets are associated with (public) blockchains and often bear (with different levels of success) the attributes of money:

– unit of account;

– means of exchange;

– store of value.

One should note in most cases, not all the three characteristics are fully met. Also, crypto-assets or tokens are not always money and have different legal classifications according to their usage and the jurisdiction.

4.2.3. *Proof*

Blockchain operation implies timestamping and can be used to create some public registries of financial transaction (e.g. Bitcoin), land registries (Bitland in Ghana) or any other kind of data.

We are even starting to witness countries such as China or France accepting blockchain record as legally accepted evidence.

4.2.4. *Autonomy*

With autonomous consensus mechanisms, blockchain are autonomous networks that (should) operate without human intervention. This ability is a key characteristic in handling a network of devices (IoT).

4.2.5. *Contract*

Even though the software code is not a contract per se, blockchain deterministic automats prevent discretionary manipulations and can thus be described as “contracts”. As implemented by the Slock.it startup, a smart lock can automate an Airbnb rental by defining a simple contract.

4.2.6. *Traceability*

Blockchain data is a decentralized and immutably timestamped ledger that offers (implicitly) a high level of traceability.

4.3. The empowerments of ME

4.3.1. *The Decentralized Finance (DeFi) movement – financial empowerment*

The Decentralized Finance ecosystem is flourishing. New services are emerging, such as stablecoins (a new cryptocurrency class which, without its usual price volatility, is generally backed by a reserve asset), decentralized exchanges, payment networks, decentralized lending, peer-to-peer insurance platforms, investment and more.

Technically, Decentralized Finance (DeFi) refers to conventional financial platforms and services built on top of open blockchain networks (to date, mostly Ethereum-based). The potential impacts (and empowerment) of this difference are:

– Wider access and reach of financial services: anyone with an Internet connection or a smartphone can access a financial service such as lending or investing. In the DeFi universe, a Wall Street trader has the same level of access as a farmer in South Africa.

– Affordable cross-border payments: by removing the costly middleman “tax”, DeFi allows more accessible services.

– Improved security or privacy: in DeFi, users have custody of their funds. If, however, the user fails to secure their private keys, then their wealth may be at risk. From a hacker’s point of view, the wealth is harder to steal since they have to break as many keys as users, rather than break the security of a single custodian.

– Censorship resistant transactions: no central entity (government, banks) can reverse or invalidate a DeFi transaction.

– A true global system: with the emergence of blockchain phones (from the high-end Samsung Galaxy S10 to the low-end Electroneum M1) and blockchain browsers (often described as a Web3 enabled browser), more and more people from different countries will be connected seamlessly, evading the constraints imposed by local centralized middlemen. This way, a Chinese farmer could seek a loan from a Canadian citizen. They would also be able to invest in a Brazilian business to pay off their debt and purchase a new tractor.

In fact, *DeFi is creating the financial system that was always meant to be.* Let's take a deep-dive with a few examples in the next section.

4.3.1.1. *Cryptocurrencies and stablecoins*

> We need banking but we don't need banks anymore. Do you think someday we can open bank account or ask for loan without physically have to come to the bank? (Bill Gates in his 2015 annual letter [AMB 15])

As discussed previously, Bitcoin is a tool of financial empowerment and is a candidate for being the base layer for an alternative financial system giving back the power to the end user and the capacity to Bank the Unbanked by disintermediating banks altogether.

Banking the Unbanked means offering digital money to the 1.5 billion people around the world (over 21% of the world's population) that cannot prove their identity and are subsequently denied access to fiat money via a bank account. The majority of these people live in Asia and Africa and are cut off from accessing basic services and economic rights, and can only transact through cash. For this kind of population, Bitcoin is starting to be used regardless of its volatility because it is comparable to the national currency volatility anyway. For instance, South Sudan's inflation was around 102% between September 2016 and September 2017, according to the World Bank. Other African countries are experiencing double-digit inflation rates (Botswana, Ghana, Mozambique, Nigeria, Zambia and Zimbabwe). It thus comes as no surprise that some of these countries are among the main Bitcoin economies in Africa (according to gobitcoin.io, a website dedicated to Bitcoin news in Africa). Still, adoption remains very low because of volatility and poor user experience with current software.

Many other cryptocurrencies, with their own set of idiosyncrasies, have followed in Bitcoin's footsteps to empower users to be their own bank and to transact in a permission-less way:

– Privacy oriented cryptocurrency like Dash, Monero or ZCash.

– Stablecoins like Tether, USD, DAI (by MakerDAO) or the upcoming Libra initiated by Facebook (to be launched mid-2020) which target the creation of a digital currency empowering billions of users to transact more easily. One should note that venture capital firms such as Andreessen Horowitz are betting on Libra, as well as other DeFi projects, including MakerDAO and Compound Labs.

4.3.1.2. *Decentralized exchanges and liquidity*

Decentralized exchanges entered the spotlight following the many issues and numerous hacks observed with the centralized exchanges. Decentralized exchanges are built on top of a public blockchain and are relying on a smart contract rather than a centralized party. New solutions like Uniswap, AirSwap or Kyber Network are emerging and empower users to trade while keeping control of their funds.

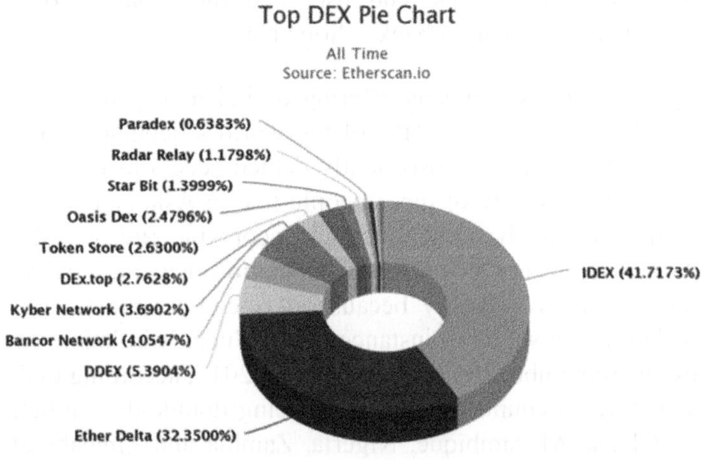

Figure 4.1. *Top decentralized exchanges by number of transactions. For a color version of this figure, see www.iste.co.uk/duvaut/IAM.zip*

According to the Etherscan DEX Tracker, these decentralized exchanges have processed, to date, more than 10 million transactions[1], mostly through the following services: IDEX, Ether Delta, DDEX, Bancor Network and Kyber Network.

This activity remains very low compared to the centralized exchanges volumes and capabilities but are significant nonetheless. This situation is due to the fact that decentralized exchanges face specific challenges such as:

– Slow order matching: the block time and sometimes the multiple confirmations need to validate a blockchain transaction which can lower the order matching time.

– Transaction cost: each trade needs to be stored on the blockchain and requires a fee. High frequency trading is prohibitively expensive!

– Insufficient liquidity: the comparatively poor user experience (as compared to centralized exchanges) prevent further attraction of funds and with it, the required liquidity to attract professional traders.

4.3.1.3. *Decentralized investments or the disruption of venture capital*

The concept of a decentralized investment (open community based funding) later known as ICO (Initial Coin Offering) was firstly formalized in early 2012 with the MasterCoin project, headed by its inventor and chief architect, J.R. Willett, describing its principles in "The Second Bitcoin Whitepaper" and launching the "fundraiser" via the Bitcointalk forum. As stated, its principle is the following [WIL 12]:

> Initial distribution of MasterCoins will essentially be a fundraiser for the trusted entity to provide money to pay developers to write the software which fully implements the protocol. The distribution is very simple, and will proceed as follows:
>
> 1. The trusted entity publishes the Bitcoin addresses to be used for collecting Bitcoins as MasterCoins are sold to the public.
>
> 2. The trusted entity publishes the final date at which MasterCoins may initially be purchased (i.e. July 31st, 2012).

1 https://etherscan.io/stat/dextracker?range=1.

3. Anyone sending Bitcoins to those addresses before the specified deadline is recognized by the protocol as owning an equivalent number of MasterCoins.

The fundraiser's objective was to fund an open source project via swapping BTC (Bitcoin) tokens with the newly minted MSC (MasterCoin). The outcome of the funding was significant as the Mastercoin project collected in one month 5120 BTC which represented enough funds to finance the development of the target software.

This funding campaign, the first of its kind, was a true innovation for 3 reasons:

– It created a new venue for funding open source projects that have no realistic funding alternative in the traditional venture captial world.

– Because the token issuance included distribution of a reserve for core developers, it introduced an economic incentive for developers to continue maintaining the software even after the funds were fully allocated. Subsequently, and for the first time in open source software history, coding and maintaining open source software were not only based on volunteering and good will, but also based on economic incentives.

– Democratization of investment (more details will follow).

Unsurprisingly, MasterCoin (later renamed Omni) inspired many other projects, notably Vitalik Buterin who was working on some Bitcoin improvements initially described in his proposal "Ultimate Scripting: A Platform for Generalized Financial Contracts on Mastercoin"[2]. Faced with the Bitcoin developer community's (correct) objections that the proposed upgrades would introduce potential security breaches, Vitalik eventually launched a new network called Ethereum, funding it through an "Ether sale". By august 2014, the Ethereum Foundation had received 31.529 BTC (Bitcoin) in exchange for selling around 60M ETH, which represented at this time around 18.4M$.

Later on and after the invention and popularization of the ERC-20 Ethereum token standard, 2017 and 2018 witnessed an explosion of Initial Token Offerings (ICOs), the new name for describing the token sale

2 https://web.archive.org/web/20150627031414/http://vbuterin.com/ultimatescripting.html.

principle. Many reasons can explain this ICO boom that we experienced and still are experiencing:

– The ERC-20 lowered the barrier to new token generation. The ecosystem around the ERC-20 token enables any newly emitted token to be stored in wallets and transacted through exchanges or decentralized applications.

– The ETH price spike created a sense of "free money" and a wealth effect for many crypto holders who could then invest in all sorts of new exciting projects, often with a fantastic narrative.

Unfortunately, the poor quality of some projects launched too early before any market-fit phase, scams and the lack of expertise of non-professional investors have slowed down the 2017/2018 boom. The figures remain very impressive nonetheless.

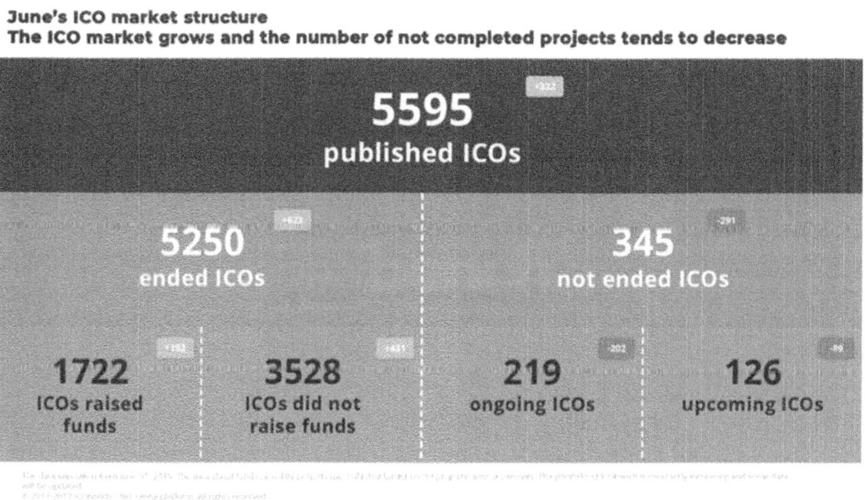

Table 4.3. *ICOs by the numbers (source: ICObench)*

Until June 2019, 5595 ICOs have been identified by the ICObench service with 1722 reaching their funding objectives or at least the defined soft-cap. Those ICOs raised a cumulative amount of $20B largely surpassing the capital injected by the venture capitalist in the blockchain startups.

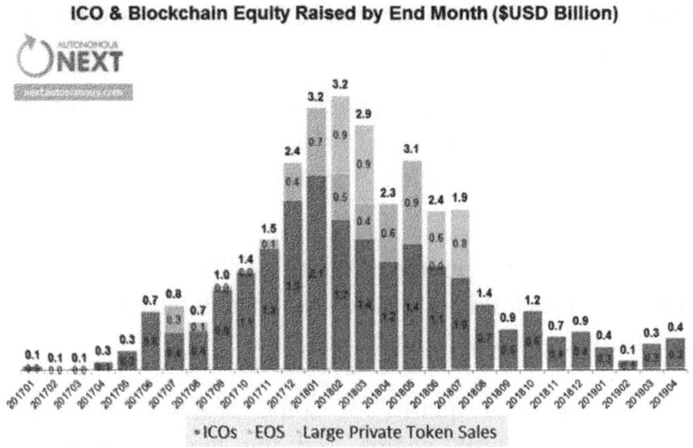

Figure 4.2. *Raised funds in ICO with EOS ICO specific display (source: Autonomous Next)*

It seems that the next phase of decentralized funding will consist of:

– *IEO (Initial Exchange Offering)*: to enhance the level of trust and liquidity of a new token, the token issuance is directly conducted on an exchange platform. To date, IEOs are gaining popularity in the crypto community. However its relative success still lacks consistency.

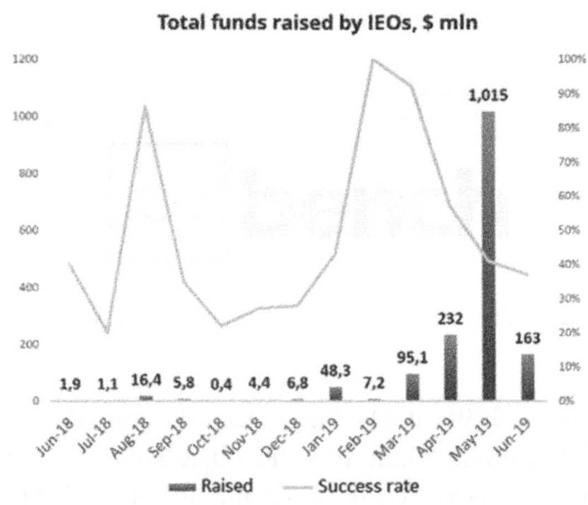

Figure 4.3. *Raised amount via IEO*

– *STO (Security Token Offering)*: a regulated financial security, issued as a token, where the record of ownership is thus kept on a blockchain. With traditional stock shares, ownership information is written on a document or recorded by some depository institution and documents issued often as a digital certificate (e.g. a PDF). The process is similar with STOs except that all proofs are recorded on-chain. STOs can be seen as a hybrid approach between cryptocurrency ICOs and the more traditional initial public offering (IPO) because of its overlap with both of these methods of investment fundraising.

Compared to ICOs, STOs, as regulated financial instruments, are perceived as less risky thanks to legislation and regulatory bodies that enforce transparency and accountability. A security token will also be backed by a real-world asset. It is therefore easier to assess whether the price of the token is fair in relation to the underlying asset.

Also, compared to traditional Initial Public Offerings IoT, an STO is cheaper because of the disintermediation of middlemen, such as banks and brokerages. Smart contracts reduce the reliance on lawyers, while the blockchain reduces the need for paperwork. This makes the whole process not only cheaper, but also faster.

Finally, fractional ownership and the ability to trade 24/7 is bringing additional liquidity to the market, especially with traditionally illiquid assets, such as scarce paintings, property and collectibles.

It is still early days for the STO market and many token issuance platforms are emerging and competing: Ethereum EIP1400, Polymath (ST20), Open Finance Network (S3), Harbor (R-Token), Securitize (DS). Raised amounts are still not significant enough, but this market is definitely to be monitored.

An interesting STO case to mention is *Bitbond*, the first legally compliant STO in Germany. Bitbond is the first cryptocurrency based lending platform for business loans that operates globally. Bitbond launched in April 2019 an STO in order to gather funds for loans, promising to share the loan interest with STO participants. The prospectus filled by the company was approved by Bafin (German financial markets regulator, the equivalent of the American SEC), which also makes it compliant with the European Prospectus Directive. At the time of writing of this document, the issued token (BB1) was about to be tradable on the Stellar DEX, the decentralized exchange available on the Stellar blockchain that has been used to issue the BB1 token.

At this stage, we have witnessed only a few regulated STO like Bitbond, but new regulatory frameworks like the one defined by the "*Loi PACTE*" in France are paving the way for upcoming European STO(s) and further financial innovation.

4.3.1.4. *Insurance*

The insurance industry may suffer from a perception issue, with a lack of transparency and misaligned incentives, since insurance companies may tend to minimize payouts and maximize their profits to the detriment of their customers. Decentralization can be a way to bring back trust. The decentralized insurance, sometimes referred to as "crowd-surance", is the act of people uniting to gather funds to guarantee compensation in the event of an unexpected loss from a community member. Compared to the known model, there is no insurance: the community is empowering itself to insure all kinds of events, and the process is being managed by the community through software and voting amongst the participants.

This fantastic proposition comes with a number of challenges:

– How can we avoid fraud? In such a (decentralized) process, fraud can be easier, especially because the "experts" validating the compensations are not always professional insurers. Also because the membership fees are hard to evaluate.

– How can we involve and incentivize the community members to participate in the day-to-day decisions and votes in the claim procedures? One of the challenge is to avoid asking everyone to participate in all votes, in which case most of the people will stop contributing rapidly.

– How can we price such an insurance product?

Among the hundreds of startups exploring this path, *Dynamis* (unemployment insurance) is exploring the solution of using the social networking data and validation points to verify a claimant's employment status among peers and colleagues. Still, the whole solution is relying on smart contracts for automating claims and peers voting on the claims.

Another noticeable startup is Etherisc, which is building a Generic Insurance Framework empowering anyone to build a parametric insurance. This solution is described by the team as a kind of "Linux" (base component) for building any insurance. Already, a Flight delay DApp (decentralized app) with hundreds of payouts is in production and the future is still to come.

Last but not least, decentralized insurance is not just the remit of startups. Big insurance companies are exploring the subject through the Blockchain Insurance Industry Initiative (B3i), gathering more than 40 industry leading companies contributing as shareholders, customers or community members. The vision and goals remain the same: it's about using the technology to deliver better solutions for end consumers through faster access to insurance and cutting out the red tape!

4.3.2. *Decentralized computing*

Could blockchain power a decentralized web that would not rely, as it does today, on only a handful of companies operating it? Various startups are working on this vision of a decentralized network and data storage that could power the next generation of the Internet, often referred to as the "Web3".

Tim Berners-Lee, who heavily contributed to the Internet as we know it, has been working on a solution to most problems he sees in today's centralized Internet. Tim has initiated the "Solid" platform, which is using the existing web protocols while allowing and empowering users to control their own data. For instance, a user moving from a web search engine to a social media platform would use its single, self-stored Solid "data pod" to verify transactions within each application. In all situations, when the user is prompted, he/she would reveal only the data needed to perform the desired activity (e.g. making an online purchase).

This vision is also materialized by Blockstack which is building a full stack software kit that abstracts away the blockchain complexity so that developers can focus on building apps on top of decentralized storage and using blockchain authentication. End-users are empowered to use applications that may seem similar to existing ones, but that are different because users retain control on their data.

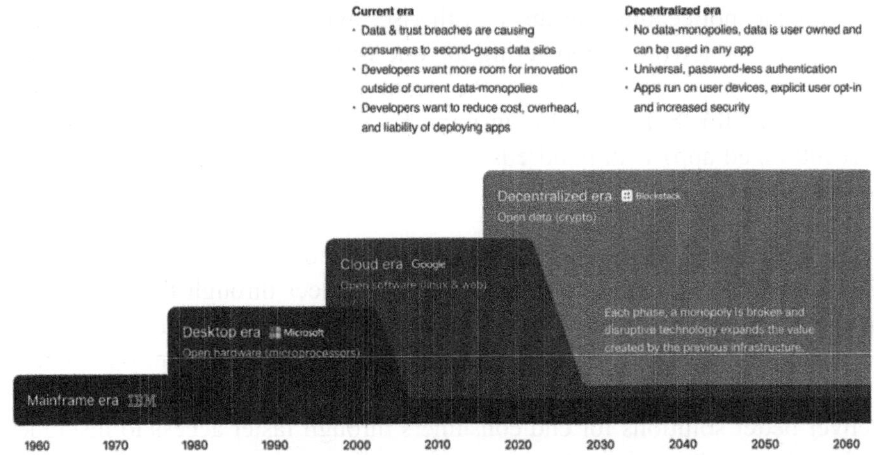

Figure 4.4. *The next era of computing (source: Blockstack presentation)*

However, this vision is not yet fully deployed, mainly because of current limited scalability of blockchain infrastructures. This situation will definitely evolve over the next months and years.

4.3.3. *Energy*

Blockchains and decentralization have also drawn a considerable interest from energy supply firms, startups, technology developers, and even national governments. This interest in the energy "blockchain revolution" has two main catalysts:

1) The rise of renewable energies: by nature, renewable energy production is decentralized. For instance, solar energy cannot be produced in a single place: solar panels must cover all the territory and cannot be managed by one and only one historic company. The blockchain can be a perfect tool for safely coordinating several companies or individuals producing energy.

2) The new electric mobility: electric scooters, cars and more are introducing new challenges to the electric grid with the energy consumption becoming highly mobile.

As energy production and consumption become increasingly decentralized, blockchain technology might become a key component in enabling the collaboration and participation of more actors trying to meet the challenges of satisfying the end-consumers in mobility.

There are multiple blockchain initiatives in the market and we may classify the use cases in one of the three following categories:

– Use cases leveraging cryptocurrency transactions or timestamping: for example, the South African startup Bankymoon has launched a Bitcoin-funded smart energy meter. Rather than settling an invoice *ex-post,* consumers are able to top-up (remotely) a smart prepaid meter by using digital currencies, thus avoiding late payments and reducing debt to nearly zero. This benefits both consumers who can pay for what they need, and energy suppliers who are getting paid on time and avoid disputes with customers and municipalities.

– Use cases leveraging smart contracts: for example, the Share and Charge initiative by the German energy supplier RWE has the ambition to enable everyone to share its charging station with others and jointly create the future of e-mobility. The end-user experience is simple and transparent since a mobile app enables one to find the nearest community charging point and displays the rates proposed by the charging point owner. Using the public blockchain, a community-based national charging point network is created and everyone can participate in it and even earn money.

– Use cases leveraging decentralized apps like marketplaces or funding: for example, the Brooklyn Microgrid by *Transactive Grid* (joint-venture between Consensys and LO3 Energy) is an experiment where some households producing surplus energy are able to sell it to neighbors. The blockchain is used as a clearing and settlement component to enable neighbors to interact without any middleman fees.

To conclude, similarly to other markets, the energy suppliers are working on energy standards and have created the Energy Web foundation, a non-profit organization focused on accelerating blockchain technology use cases across the energy sector. Below are some quotes Hervé Touati, a managing director at RMI and president of Energy Web Foundation, to better demonstrate the challenges and the value added by the blockchain for this industry [ZER 17]:

> The main challenge of the electricity sector in the 21st Century is to *integrate more renewable energy into the grid in a cost-effective fashion* in a context of largely flat or diminishing demand. The only way we know how to do this is by automating the demand side by allowing many more participants in the grid. That means automation at the distribution edge, and integration of this automation with wholesale markets.
>
> [...]
>
> We are excited by the potential of blockchain technology as an enabler to realize that vision. Blockchain will not be the only building block of the 21st Century grid, but it will most likely be a key building block. It also provides much higher levels of cybersecurity essentially for free – which addresses, as a by-product, one of the key concerns of utility executives when it comes to distributed energy resources.

4.3.4. *Booking services*

In most booking platforms, centralization has led to an imbalanced situation, where online platforms have gained increasing power over the actual service providers who bear the operational costs and investments. This situation exists in many industries such as hotels, car rental, shows and events, restaurants… and many other industries that rely on digital distribution. The network effects of such platforms are preventing existing players from operating in a free market.

With blockchain networks emerging as a new global infrastructure, there is an opportunity to create vastly different power structures. The peer-to-peer capability of the blockchain may enable the capacity for service providers to get closer to their customers, creating a leveled playing field for them to take back control of their digital distribution.

Many blockchain startups are trying to break the monopoly of online booking platforms. BTU Protocol may be one of the most subtle and valuable tokens in this category: unlike many other approaches, BTU operates at a protocol layer and it's up to service providers to build the end-consumer applications. This may capture more value according to the fat protocol layer thesis developed by Union Square Ventures and updated recently with the "Thin Applications" article[3]: the market cap (value) of a protocol layer always grows faster than the combined value of the applications built on top, since the success of the application layer drives further usage and speculation at the protocol layer.

BTU Protocol is indeed financial technology (fintech) allowing enterprises to build new distribution channels with controlled margins for promoting their services. The solution includes the software and the BTU token that is used to reward and incentivize direct business. The solution is built on the following principles:

– 0% commission for using the software. This is possible because the value goes into the BTU token whenever it is used. The BTU token can be compared to an atomic unit of value of the market handled by the BTU software. This situation also creates an incentive for the first users of the system that create the initial value. This main difference with existing marketplaces can participate in boosting the ecosystem birth and its usage.

– Companies are paying BTU rewards to end users or to medias and websites promoting the services. This way, enterprises are building new direct (and semi-direct) distribution channels by deciding and controlling the user cost acquisition. This is in contrast with the usual situation where margin and user cost is determined by a broker bringing (or not) most customers.

Software companies from the hotel industry have already been empowered by BTU Protocol to build the "BTU Hotel" services which allow them to control the margin for this channel, while making happy customers or partners as they benefit from a 5% reward for using the service.

3 https://www.usv.com/writing/2016/08/fat-protocols/ and https://www.placeholder.vc/blog/2020/1/30/thin-applications.

This logic is currently being deployed in new industries as activities, car rental, cruises and more.

What are blockchains actually good for?

Blockchain networks are useful when it comes to coordinating a large number of players and disintermediating a trusted middleman.

Before using a blockchain, one should ask the following questions:

– Is immutability or censorship resistance needed for the use case? Immutability is the assurance that no ecosystem participant can ever change the records. If immutability is not needed, a traditional database can do the job.

– Are many entities involved or are we solving a problem "just" inside a single organization? Since trust is achieved inside one organization, blockchain makes sense only when many entities are involved.

– Do the entities involved trust each other? Blockchain removes the need for trust between the participants, but if trust already exists between the participants, a blockchain network is overkill.

– What are the performance and scalability requirements? If the performance target is too high, the current performance of blockchain may prove to be insufficient.

Empowerment

Whatever the industry mentioned in this chapter, the purpose of blockchain is to shift power away from platforms and middlemen and to empower end-users. This empowerment comes with multiple (positive) consequences:

– Innovation: innovation is unleashed by removing the permission barrier and empowering the citizens and developers.

– Power experiments: with new governance models and abolition of hierarchical structures, new socio-economic models are being experimented.

– Funding: new funding models enable new categories of projects that would never have emerged otherwise.

New asset class and flows of value

With the empowerment of end-users arises the question of value creation and how it is redistributed to participants. In most cases mentioned in this blockchain chapter, the value flow is transferred via a digital token often issued with a limited supply.

Unlike in traditional companies, value creation is not captured by the company and distributed to the equity holders, the value created is distributed to the users of the blockchain protocols holding the protocol tokens. This paradigm shift drives user adoption with an incentive to contribute to own their wealth.

Box 4.1. *Takeaways*

4.4. Glossary

BTC: ticker symbol for Bitcoin currency.

DeFi: Decentralized Finance. Finance automated on top of smart contracts and operating without requiring traditional actors.

ICO: Initial Coin Offering. Token sale for funding a new token-based project.

IEO: Initial Exchange Offering. Token sale similar to ICO but conducted via an exchange company.

IoT: Internet of Things. Network of devices over Internet protocols.

PoW: Proof of Work. The algorithm used, for example, in the Bitcoin network to reach a consensus on the state of the valid transactions known by the network.

STO: Security Token Offering. Offering that involves a tokenized security product (financial).

5

Internet of Augmented Industry

"Innovation is the market introduction of a technical or organisational novelty, not just its invention."
Joseph A. Schumpeter

5.1. Formation of a systematic approach to implement digitalization in production

5.1.1. *Organizational foundation*

The aim of this chapter is to provide some examples on the application of digital tools and augmented reality in the production areas of aircrafts and spacecrafts at Airbus. To achieve maximum benefits, we believe in the importance of having one standard of these tools for the whole production perimeter, even if the final application may vary to some degree, taking into account the respective business needs and constraints. It is necessary to understand some background of the organizational setup of the business described before implementing standardized processes.

In 2017, a decision was taken to reorganize the whole production perimeter, which resulted mainly in building two organizational units: One representing the aircraft final assembly lines, including maintenance, repair and overhaul activities, and the other grouping together the aero- and space-structure activities and the assembly facilities for spacecrafts. It should be noted that the first step was taken in 2014 when aircraft production areas were aligned. Therefore, this second step was a logical next step. However, all of these production areas have, in the past, hardly been connected to each other, which resulted in different processes, methods, tools and even mindsets in running their respective production areas. In addition, it has to be mentioned that each of these areas has a long history and ran the respective kind of businesses in a very particular way: some producing for institutional business, some working within governmental contracts and some within commercial contracts, which have been won in open competition on the world market, even producing parts for competitors – with a span between a large number output regarding this specific kind of business and building prototypes and one-offs, but always at the leading edge of technology.

5.1.2. *Selection of digital initiatives*

As mentioned, a very scattered and diverse landscape was the starting point. Reorganization decisions were taken in order to harmonize the whole production floor, regarding processes, methods and tools, despite knowing it would take a long time to achieve. However, even with the very diverse business, overall benefits were seen as being convincing enough to do so. This decision came with the strong guideline to take advantage of the digitalization of production; to modernize it and to foster the harmonization even further. Based on this, the decision has been made to carry out – with external support – a collection of different digital tools and methods in use and in development during 2017. This turned out to be a very complex process for several reasons: quite frequently the same process was tackled with different digital tools, single process developments have been supported by external institutions/authorities, business cases were based on singular or different calculation schemes, some in an early stage of development, and many more reasons besides. Finally, close to 200 processes/tools were counted and needed to be taken into account. After collection, a process started to drive down the number of initiatives to be applied in general for all production activities – this was based on business cases for the overall production perimeter, operational needs, contribution to harmonization, current state of development/implementation, obligations with partners, to name but a few. Needless to say this was a lengthy process which finally ended by downsizing to 20 prioritized initiatives, of which 12 were applied locally and eight were initiatives, which fully supported the transversal criteria to apply them across the whole scope of the production perimeter. An agreement was found to apply the eight in general, but to follow the other 12 on a local basis, supported by single business cases.

5.1.3. *Implementation of digital initiatives*

As the agreement on the most valuable initiatives proved challenging, the implementation phase was expected to be even more critical. Each initiative had a specific business case and received a respective implementation road map. To support fast implementation of the eight transversal initiatives, it was decided it would be supported in a particular way by cross-functional implementation teams, each with a dedicated management sponsor. The team and the sponsor had transversal end-to-end responsibility for the implementation and achievement of the respective initiative. Additionally, it was decided it would be implemented in three waves, providing an opportunity for quick-win and fail-fast and, from this, to use spin-offs for the next areas to come. Common for all initiatives was a monthly status report which focused – besides the progress of the implementation of the initiative – on the progress regarding the achievement of the respective business case.

5.1.4. *Selected digital initiatives*

The digital initiatives chosen can be grouped as follows:

– planning and controlling of production;

– robotics;

– augmented reality.

In this chapter, we will briefly provide some ideas on what has been done regarding planning and controlling, and explain in more detail what has been done with regard to robotics and augmented reality, along with some use cases.

5.2. Digital planning and controlling of production

In the aerospace sector (as in the majority of industries), product lifecycle management (PLM) starts with a feasibility phase, followed by a conceptual definition phase and finally, the development and serial production phase – three phases which take decades from the feasibility up to the last manufactured serial aircraft. Product development runs from feasibility studies up to the first serial production – right after prototyping for tests and certification. The transition between product development and product manufacturing is called industrialization. Its activities are day by day starting earlier into the PLM scope and starting late at the definition phase, covering all development and reaching the early serial production phase. Manufacturing a product in aerospace sectors reaches the final stage at the development phase and the complete serial production phase.

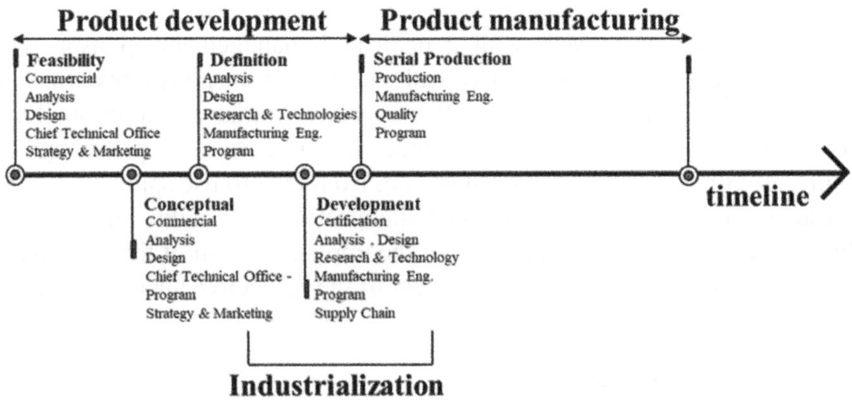

Figure 5.1. *Product lifecycle phases*

Following this rationale from a data point of view, customer requirements captured in the feasibility phase are translated into product requirements at the conceptual definition phase, in which industrial requirements are also taken into account. The output from the definition phase is a product, digitally defined from the design and industrialization points of view. Data coming from simulations and industrial plans are gathered into the development phase to generate detailed build processes, plans and work orders. Data contained on this information will be exploited intensively at the serial production phase, executing defined processes. To effectively execute those processes, planning all activities with real data from the shop floor, controlling the progress of each of them and managing disruptions in case they occur, are some of the core activities at the production phase. This journey of data from customer requirements up to execution and disruption management is usually known as the end-to-end (E2E) product lifecycle.

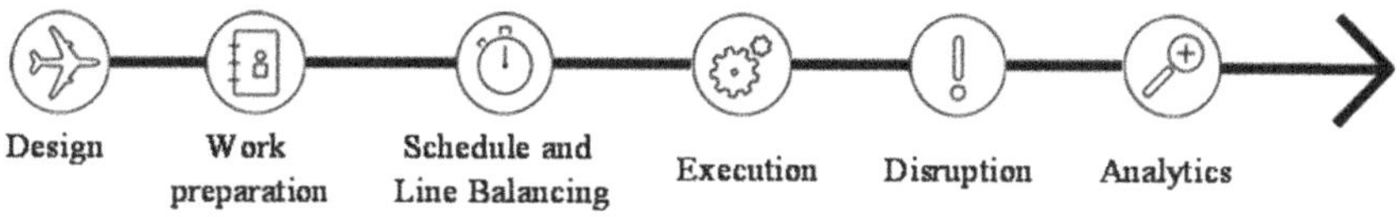

Figure 5.2. *End-to-end product lifecycle activities from a data point of view. Digital initiatives usually impact one or more of them*

Digital initiatives, with regard to planning and controlling of production, were selected as one of the most value-added factors to boost the digital transformation at operations level. They mainly affect schedule and line balancing, execution and disruption management. Any digital transformation faces multiple technological challenges. In particular, there are some technologies that are extensively used at industry level that are related to smart devices and smart tools, data exploitation and visualization, robotics, end-to-end digital continuity and IT infrastructure. From this perspective, digital planning and controlling of production mainly involves data exploitation, end-to-end digital continuity and IT infrastructure topics. For each of the following use cases, this perspective is detailed in the corresponding explanation.

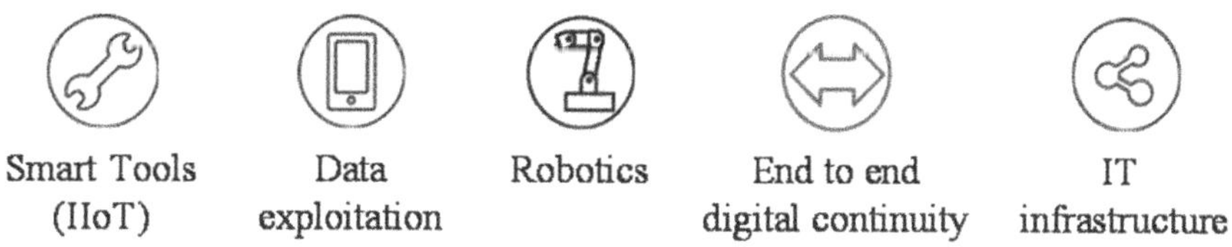

Figure 5.3. *Industry 4.0 technologies addressed in the digital initiatives implementation plan*

5.2.1. *Digital planning preparation*

One of the key activities for manufacturing engineering and production control areas is to plan the sequence for manufacturing and assembly operations, starting from product design and the bill of materials (BOM). It is performed at the industrialization phase, according to section 5.2 on schedule and line balancing activity. An accurate schedule and resource planning is a key enabler for on-time and on-cost delivery. Traditionally, it is done by relying mainly on expert knowledge familiarized to perform complex tradeoffs between competing priorities, shared resources, parts supply and other specific constraints, resulting in a very time-consuming task.

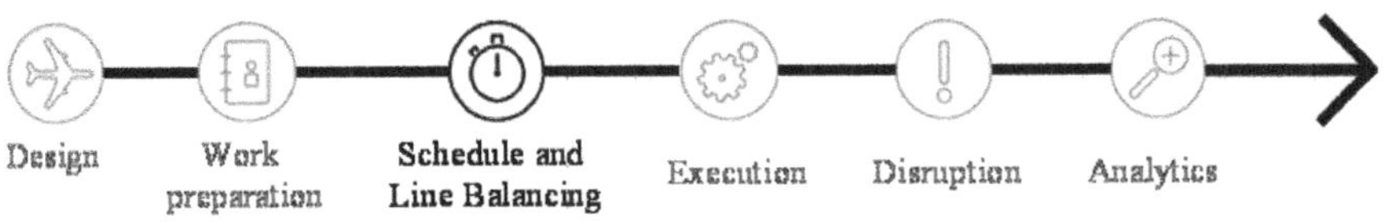

Figure 5.4. *Digital planning preparation initiative boosts data management mainly at schedule and line balancing activity*

The aerospace manufacturing environment multiplies the complexity of such activity due to different precedence constraints usually converging into one single production line. As an order of magnitude, one single station in a final assembly line could group tens of thousands of operations, with several thousands of precedencies among them. The amount of layers within the manufacturing structure (from high-level build process down to detailed part level), and the iterations to be performed in the planning between first article and serial production, are two particularities within aero and space structures that make this activity even harder. In addition to these challenges, any digital solution should cope with continuity of the manufacturing data along the process, which requires interfacing with enterprise resource planning (ERP) and IT ecosystems to guarantee the absence of manual intervention through the data journey, across the process. Even more complex, as long as the digital mock up (DMU) contains the core information on aircraft or spacecraft design and details of the bill of materials (BOM), any digital solution should also interface DMU and BOM to guarantee the afore mentioned digital continuity of data. As an indication of this complexity, the number of parts in a single-aisle aircraft can be counted – usually about half a million!

The solution adopted consists of an advanced digital planning and scheduling tool, fully interfaced end-to-end into manufacturing systems. This helps the engineers and production controllers to develop a detailed production schedule, given a number of operations, process constraints and resource constraints, using a set of objective functions (e.g. better lead time or higher line saturation) and also calculating the resulting critical path. This solution reduces the time needed for initial planning and scheduling, under different scenarios, from a number of working days down to a matter of hours. In order to better adapt the different areas within operations, maintaining a harmonized and simple structure, all impacted jobs were grouped into two different entry types: one for standard production baseline schedule calculations, and another for daily incidence and inflow impact on schedules.

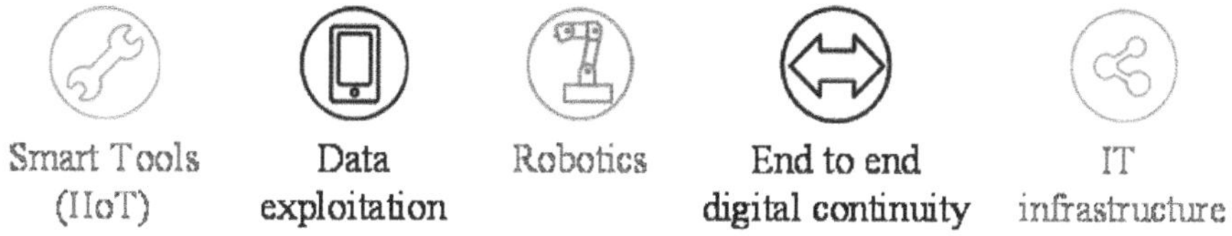

Figure 5.5. *Challenges facing digital planning preparation initiative in data exploitation and end-to-end digital continuity technologies*

During the development of the planning tool, the creation and fine-tuning of specific algorithms, centered in the management of complex precedencies and shared resources, was one of the most difficult challenges. In order to manage the information needed for these algorithms to run, special high-speed computing issues were also faced. During the massive implementation of the tool, two main challenges were faced with regard to change management and data continuity. Due to the complexity of precedence constraints on processes to be planned, new skills in mathematical topics needed to be put in place, training manufacturing engineers and production control engineers in new areas, not previously taken into account. This reinforces the idea that a good final solution is based on a strong precedence constraint model, which takes time to set up and requires new skills to build properly. Data continuity challenges were focused on the creation of interfaces with existing IT applications, from where most of the inputs came to where most of the outputs went.

5.2.2. *Digital production planning execution*

As mentioned, planning for a specific product is achieved in two phases. First, manufacturing engineering starts by designing data and BOM inputs and ends with a standard sequence as output. Second, production control enriches the standard plan with real resources, data and information. After these preparation tasks comes the execution phase for the plan. This phase essentially consists of assignment of workers to tasks, production progress tracking and disruption management.

Historically, work assignment and production progress tracking were done by printing huge paper sheets containing the theoretical plan, involving workers themselves in coloring it. In this way, information became obsolete and they had no visibility of the inflows or the problems of the operations. These simple planning methods cannot adequately address complex situations related to competing priorities, parts supply and other constraints, which can sometimes necessitate daily changes to schedules. Digital planning execution was used to digitalize this process, offering an interactive way to perform work assignment and station tracking. It also provides online shop floor information about the status of each task, as well as all problems that may affect its execution, being totally integrated with the manufacturing execution system (MES) and enterprise resource planning (ERP).

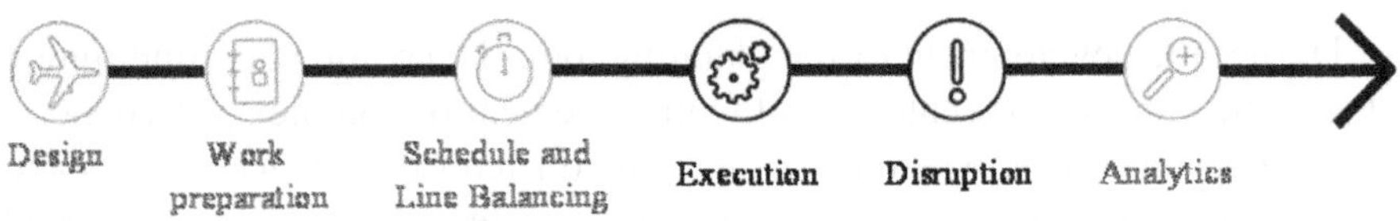

Figure 5.6. *Digital planning execution initiative impact on data managed at execution and disruption activities*

The solution was an interactive digital tool to be used at the shop floor level that allows work assignment and production tracking. It displays the planning sequence of a station and also the real-time information of the status of the tasks, making it possible to know if a task has started, is blocked or running. It also shows all relevant information about production, including incidences, missing parts or non-conformities affecting the process. It enables fine-tuning of the daily plan while visualizing the critical path, all of this coping with digital data continuity principles, which led to a massive interface with IT production control and the manufacturing engineering landscape. Apart from the benefits at administrative and reporting level, the step forward is to have a real and online way of measuring adherence to the production plan. This also allows us to mathematically detect when saturation on an assembly station has reached the point at which a new balance on the workload can be done, to free new staff.

Non-conformities, incidences when performing tasks on the production line, faults and missing parts affecting the sequence are usually referred to as disruption. Disruption management is the third area in planning and controlling of production, where it was decided to implement one digital initiative across all Airbus DS. Disruptions to the manufacturing processes are due to a vast range of causes and could affect, or not, the final delivery and quality of the product.

One single flow for incidence is a digitalization initiative aimed at providing a single interface for the worker and a clear and unified process to raise, manage and track incidents and improvements, which may enhance the process. To implement this idea, a digital tool has been developed that allows blue collars to report different types of incidents and to treat them in a more organized way.

This initiative was to enhance the management of incidents and improvements detected throughout the production process. The main benefits associated with this initiative are administrative savings and support efficiency improvement. Administrative savings includes physical displacement of blue collars to an available PC on the shop floor, user logins, applications start-up, manual typing of information, sketches and/or picture attachments, etc. Increased efficiency is achieved through mail management, searches into historic data, priorities management, data analysis storage, synergies between different supporting functions, etc.

The technological challenges for these use cases were real-time data exchange – latency of data in the IT ecosystem and interfaces between new digital tools and existing ones, visualization of information and a worker-centered interface showing, at any stage, only the relevant information required without overexposing the blue collar to additional non-value-added information. These were the two key points to manage during the development of the tool for widespread implementation.

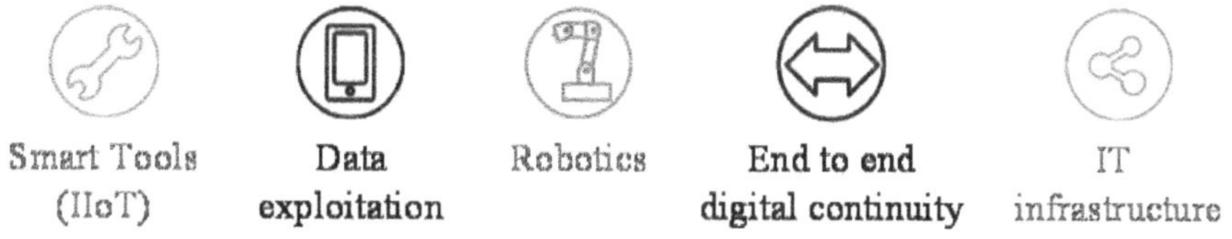

Figure 5.7. *Digital planning execution initiative faced challenges at data exploitation, end-to-end digital continuity and IT infrastructure technologies*

5.2.3. *Digital controlling of production*

At Airbus DS, operations activities rely on a lean foundation. The operational excellence system (OES) represents a step forward from the already implemented lean manufacturing and integrates different improvement initiatives of the company as well as best practice at operations level, in a coordinated and structured way. This system allows customization for the different types of businesses within manufacturing: final assembly lines, aerostructures, spacecraft and MRO.

The OES establishes a tailor-made deployment map of tools and methods based on assessments, enabling operational excellence. It takes as the input the strategy and performance of the area, as well as the current excellence system deployment status.

OES in Airbus DS identifies four pillars (people, operative model, performance improvement and customer), and in each one, a set of tools is stated, considered as essential to implement core concepts on each pillar. Those tools usually became an enhanced evolution of standard lean tools (e.g. the performance improvement pillar states SQCDP boards and production control boards as essential tools to implement performance routines). The aim of operations has been not only to master those tools but also use the opportunity of digitalization to make the implementation easier and connected.

Taking the two use cases above (SQCDP boards and production control boards), the aim of digitally controlling the production focused on digitizing currently paper-based SQCDP and production control boards, connecting them to existing IT landscape production applications. This approach was not only to foster and improve decision-making by improved transparency, but also to stream the exchange of information with the possibility of analyzing and creating graphics for pending actions, status, lead time of actions and customizable key performance indicators, using OLAP technology cubes. It is paving the way towards systematic digital-driven analytics into serial production and manufacturing.

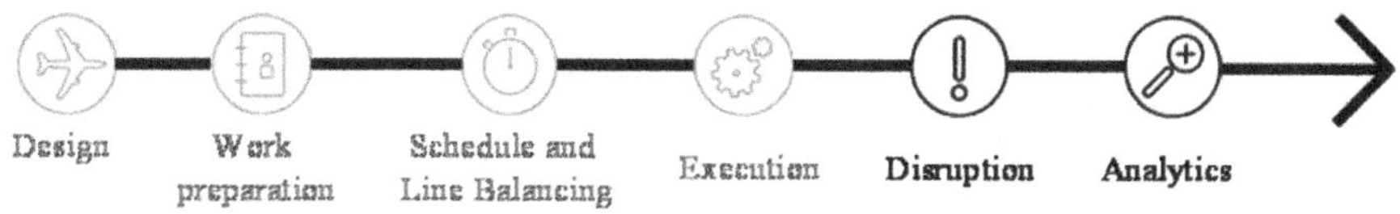

Figure 5.8. *Digital control of production initiatives managed data mainly from disruption activities and paved the way towards digital analytics*

In the use case for digitizing SQCDP boards, a key principle from the start was to foster collaboration by exchanging information between all existing SQCDP boards, and between all active areas and departments feeding those SQCDP boards regardless of the physical location. The features are attendance control, access control, historical evaluations quick view and activities management key indicators, all of this with a log of actions for all the users. The main benefits have been identified on reporting activities (in one single plant, an order of magnitude is about 50 SQCDP boards, interchanging actions, due dates and different information on a daily basis) and an average lead-time reduction of about 15% when answering actions open on digital boards, compared to paper versions.

In the use case for digitizing production control boards, the key principle from the outset was to enhance the visibility and consolidation of a set of common and harmonized key performance indicators (KPIs) on each plant, or even on each assembly line, across the Airbus DS facilities footprint. The control room is the key place to gather information of the different departments in which key topics and projects are raised and have continual follow-up by the multifunctional teams. Traditionally, production control boards were a paper-based tool (in the best cases, a stand-alone, isolated, Excel-based tool) with a slow flow of information based on paper transfer, mail management or shared folders for uploading of documents. Adding the possibility of having tables, graphics, links, URLs, images and reporting boxes of the projects, and customizing this, the user has only the information valuable for analyzing global status, including the Gant chart view and milestone view for an overview of the key milestones of the business unit's projects. The different possibilities guide the team in setting a roadmap for achieving objectives. It is the easiest way to see how effectively a function is achieving the key objectives, not only having KPIs standardized as BSC but also allowing us to create personalized ones prioritizing which information is valuable for measuring a business unit.

This high level of possible customization led to challenges on data exploitation techniques; however, the link between different boards and control rooms, as well as the interfaces with existing sources of data, were major challenges. IT infrastructure needed to be revisited to adapt to the new computing needs and data storage security.

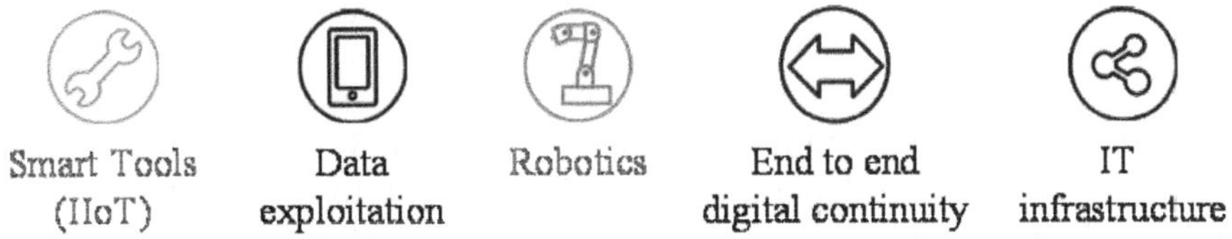

Figure 5.9. *Digital control of production initiatives faced challenges mainly at end-to-end digital continuity and IT infrastructure technologies – due mainly to data security reasons*

5.3. Robotics

Manufacturing in the aerospace sector has, in the last few decades, undergone significant automation, which began in the scope of single-parts manufacturing and has continued to extend to assembly operations and functional tests. This process did not happen at the same level as in other industries such as the automotive industry, with which it is frequently compared, but has led to a real transformation at aircraft manufacturers. In fact, a high level of automation has been achieved in single-parts manufacturing technologies (automatic lay-up operations in composite parts, numerically controlled machining of metallic parts, etc.), enabling significant cost reductions and an increase in quality in the early stages of the supply chain.

In contrast, the proportion of manual activities is still high in assembly operations, which makes the use of automation more difficult on the one hand and brings an opportunity for significant savings in assembly operations, on the other.

The origin of these savings remains in the nature of the activity to be automated and in the number of repetitions for that activity. A machine, as a physical fatigue agnostic, can develop repetitive tasks at a more stable rate than a worker. This reduces the number of failures and costs due to repairs, and increases the quality of the operation. In addition, the automation frees highly specialized aeronautics industry workers from repetitive, manual tasks, stimulating professional development.

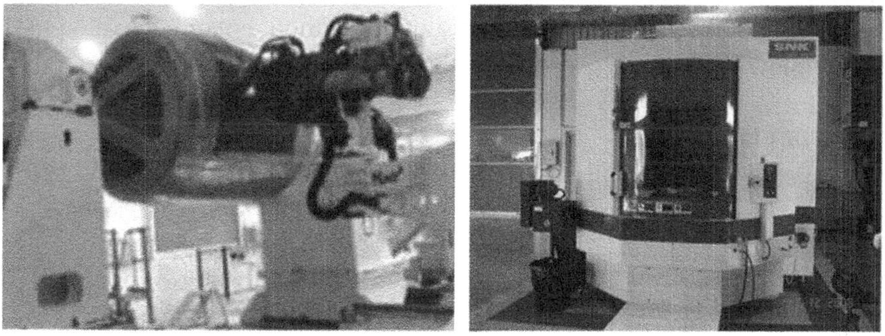

(a) (b)

Figure 5.10. *Carbon fiber automatic lay-up machine (a) and numerical control milling machine (b). For a color version of this figure, see www.iste.co.uk/duvaut/IAM.zip*

Finally, poor ergonomic situations sometimes occur in aeronautical assemblies when working in small areas, when realizing repetitive efforts or when applying certain injurious products. Automation of these operations increases the safety of workers and is the one of the most powerful reasons for studying viability of an automated application.

5.3.1. *Pin milling robot for Jo bolt rivets*

One of the most interesting challenges nowadays for airframe manufacturing is the assembly of complete, advanced-material airframe sections, built with the high accuracy specified by design offices. Composites are increasingly used for designing skins, fairings and all types of complex-geometry wet surfaces. Joining up these parts to the inner structure is still done by traditional methods based on bolts, rivets and tight diameter tolerance drilled holes. Jo bolt rivets are one of the common choices, due to the fact that installation can be done from one face of the stack, instead of from both faces, with the consequent ergonomic relief and simplification at assembly point.

After the installation of Jo bolts and rivets, a final milling operation must be performed to remove the metallic bolt pin that allows the rivet to close and join the stack (Figure 5.11). This milling operation is known in industry jargon as the "shaving" operation, and usually represents a manual and poor ergonomic process which is time-consuming, without any added value to the final product.

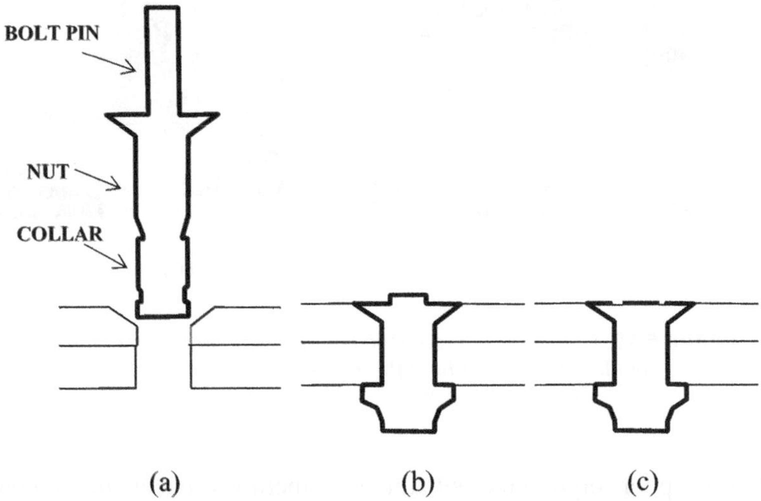

Figure 5.11. *Three-part Jo bolt description (a); Jo bolt installed before the final pin milling operation (b) and Jo bolt final installation after pin milling operation (c)*

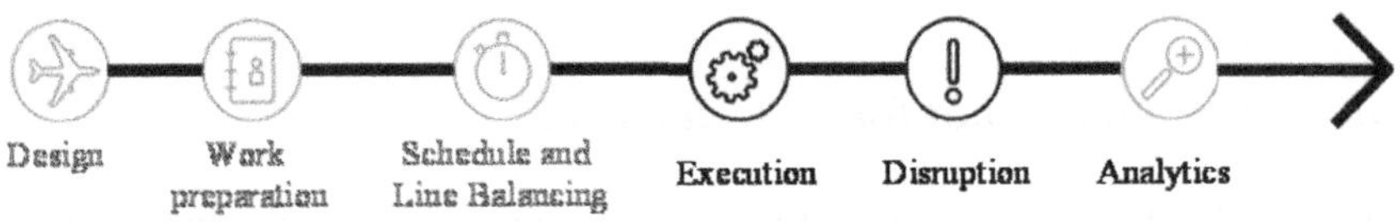

Figure 5.12. *As a robotic solution for assembly operations, the pin milling robot consumes and generates data at execution and disruption activities*

Global Industry 4.0 applications have increased the demand on robots and automated devices due to the democratization of its usage, which has turned companies away from hugely expensive, customized automation solutions towards more flexible, standardized, cheaper solutions. Little by little, the concept of huge machines designed ad hoc and dedicated to a single product is being replaced with the use of smaller, more flexible cells that do not need special foundations, and can be used for different products. Automation of Jo bolts and rivets shaving – and automation of their quality verification – was identified as one area in which those standard robotics platforms could be applied: there was no need for high-accuracy positioning, and the necessary payload to carry by the end effector was not too high. This circumstance led to the solution of a universal robot to locate the same manual shaver machine, by means of a specific pick-and-place end effector, and somehow enhanced the pin length verification tasks.

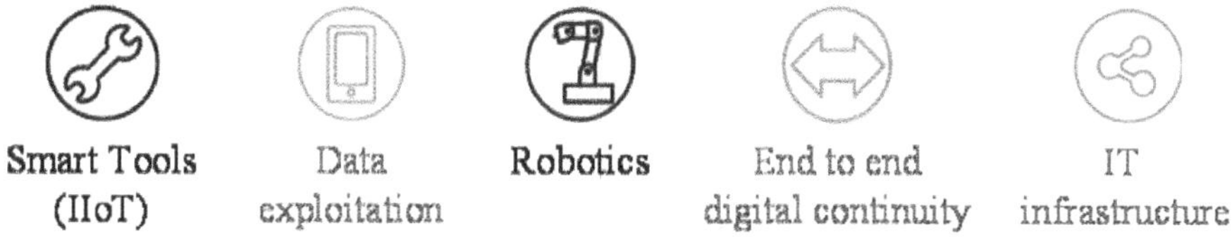

Figure 5.13. *Pin milling robot faced challenges mainly at robotics technologies and smart tools – provisions for future connections to MES*

From a technological point of view, the system architecture is a classical one with opportunities taken from the digital era. On the one hand, a low-cost end effector is based on the original idea of using exactly the same manual machine, currently used in the manufacturing process, and the integration of pneumatic air supply and chips extraction. On the other hand, an end effector positions and orients the control system to achieve the right tolerances in the final product. The system also needed a laser profile meter for checking operations. Finally, an industrial PLC-based software and hardware was needed for application control and for future connection to the MES system. Industrial Internet of Things provisions for connecting signal and sensor output to the MES system were also taken into account for the next step.

The benefits achieved after this implementation were the classical ones after process automation, including reduction of operating time and personnel, as well as consumable milling bits (hence, recurrent costs), as well as improvement in product quality, ergonomics and occupational health (hence, non-quality costs, and health and safety costs). Lastly, integration with high cadence pulse lines (hence, freeing valuable personnel for other important activities). In addition, the low-cost investment made the payback for this use case attractive.

As long as the application tried to make use of the existing manual shaving machines, one of the main challenges was to adapt the rigid and highly repetitive robotic end effector to the existing manual machines. Provision for connecting this robot to the MES in the near future was also a source of new challenges within the Industrial Internet of Things field, regarding interfaces with actual MES and ERP, as well as data protection and security protocols.

5.4. Virtual and augmented reality use cases

In recent years, virtual and augmented reality electronic devices have progressed in a manner that has triggered the possibility – in a massive way – of exploiting data stored in the IT landscape of each and every company. This is the case for data stored in the digital mock up or ERP systems: earlier, a PC or a powerful work station was necessary to properly visualize those data, and new devices today provide universal access to it by means of tablets, smartphones, tactile screens, etc. All users, involved in any stage of design, manufacturing engineering and production, can today find very specific and useful devices, particularly appropriate for their different requirements – from an industrial tablet able to show 3D work orders at shop floor level, to laser projection devices showing the right data at the right time, rendered in a holographic style, directly on top of the real world.

Two important aspects need to be taken into account for the application of these technologies in a real production environment. The first is to find the most appropriate device (even if it is not the most advanced) in each manufacturing process. The second aspect is to choose just the right data – and no more – to be delivered at every task within the process, adapted to each and every individual, if possible.

In line with the first aspect, the display of digital mock ups at shop floor level, by means of an industrial tablet is a simple way of exploiting virtual environments generated on a computer – a very crude but effective virtual reality application. In line with the second aspect, a laser mark, simply overlaid on top of real aircraft structure, is sometimes the only help an operator needs to perform the job – a straightforward augmented reality application, but a worthy one as well.

5.4.1. *Digital work order*

The work order is the essential document describing tasks and instructions in an assembly line or a manufacturing cell. This document usually calls for design documentation (in the past, drawings, and nowadays, 3D digital mockups), bill of materials, specifications, standards and special tooling for each step of every task. Non-conformities are also raised and referenced in the work order structure, as well as final disposition and product validation checks.

Each manufacturer serial number (MSN) has a unique set of work orders containing detailed information about who is performing each task, when the task was started and how long it took to complete it.

Traditionally, the work order has been a paper-based documentation set progressively fed into, consulted and checked by a collection of operational stakeholders (production control, blue collar staff, quality inspectors, manufacturing engineers, quality representatives, etc.) throughout the evolution of the manufacturing process.

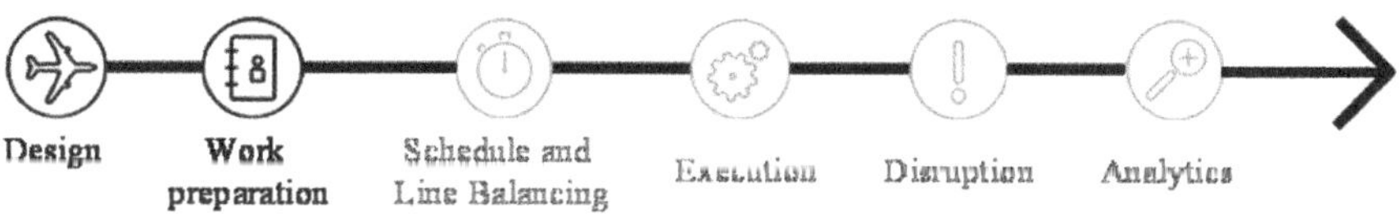

Figure 5.14. *Digital work order initiative managed data from design and work preparation activities*

The goal of this initiative was to change the type of documentation provided to the shop floor, evolving from a paper document to a paperless system. It can be distinguished into two types of digital work order depending on its content: 2D digital work orders if its content is based on 2D documentation; or 3D digital work orders if its content is based on 3D models or digital mockups.

(a) (b) (c)

Figure 5.15. *Paper work order (a); 2D digital work order on a tablet (b) and interactive 3D digital work order on a tablet (c)*

Benefits of these initiatives were clearly derived from having the right documentation, at the right place, at the right time. In fact, loops to set up the documentation have been reduced, the availability of information to the worker has improved (not so much the quantity of information, but the quality and accuracy of the requested information), non-quality costs due to documentation issues and incidents during the completion of tasks have reduced, and the cost of paper (and its management) has been drastically reduced.

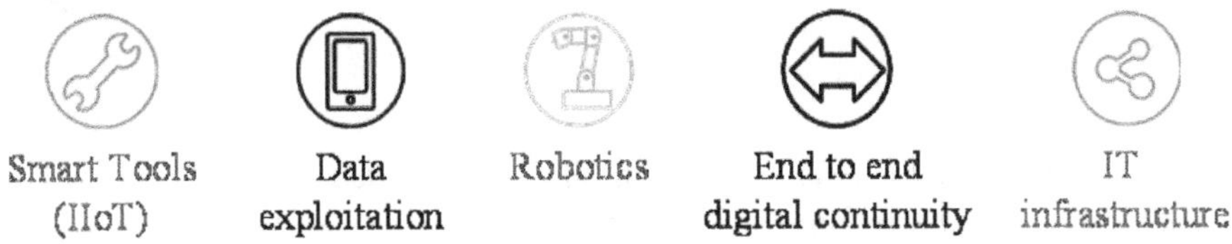

Figure 5.16. *Digital work orders faced challenges mainly at data exploitation technologies as well as end-to-end digital continuity*

From a technology point of view, visualization of information, real-time data exchange, data collection, display of visual feedback and virtual reality has been realized. Virtual reality technologies have been key for this transformation to become a reality since it was needed to find easy ways for the operator to navigate and become immersed in the digital industrial environment (product, tools, etc.). Hardware challenges related to the industrial preparation of devices (autonomy, degree of electrical protection, graphic and visual performance, etc.) and software challenges related to connection and reachability of data (interfaces, standard structure for data, graphic engines for visualization, etc.) were forced to come to an industrial solution.

5.4.2. *Remote assistance*

Whenever operators have questions during the completion of a task, those acting as support functions are traditionally contacted by phone or e-mail to provide answers. Solutions could be also provided in the same way, however physical displacement of the assistants from the shop floor area – where the question comes from – is sometimes required. Within this process, data in several formats (pictures, sketches, digital mockup snapshots, etc.) might be produced and exchanged, and complex loops of data acquisition, measurements and other requests could be required in the case of non-conformities characterization.

The remote support tool provides the ability to perform these activities more quickly and more easily by a video conference system, enhanced with additional functionalities, integrated with Airbus DS processes. It allows communication between shop floors and support teams in real time. This kind of system allows us to generate a collaborative environment between several workers, sharing audio, videos, photos, 2D and 3D mockups, etc. The objective is to resolve incidents faster and in a more efficient way by the elimination of waste and reduction of non-value-added activities; for example, movement of workers, transcribing data to upload it to the MES system, waiting time, etc. Additionally, this kind of technology allows better organization and optimization of support teams.

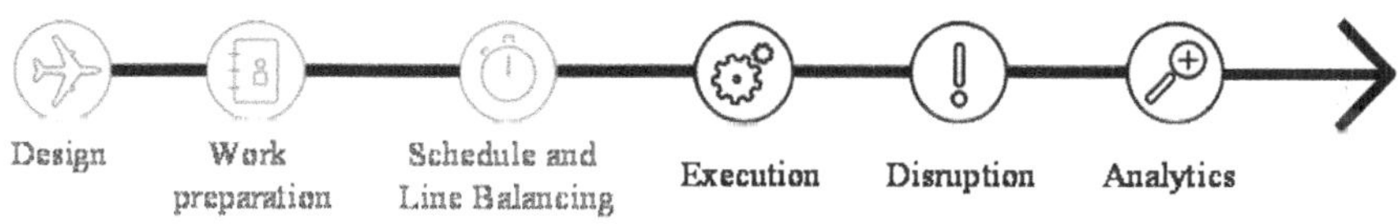

Figure 5.17. *Remote support tool affected mainly data at execution and disruption activities. Remote support is a source of information for further analytics activities*

The proposal deployed at Airbus DS is based on two complementary areas: the first focused on communication between plants and remote field teams during the testing phase of spacecraft. The second focused on direct communication between operators and support teams, triggered by production control at exactly the point in the work order at which the incidence has occured.

Technological challenges faced during deployment of these solutions were the integration of systems into a high-security IT environment, the traceability and storage of documents shared, and support request ession data (start time, who, end time, etc.) on the digital work order tool.

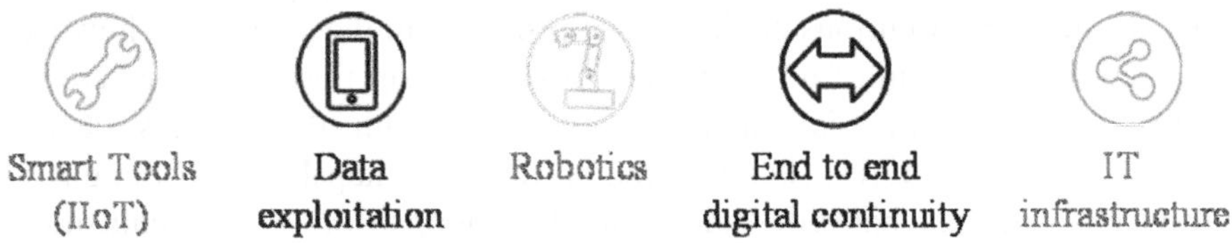

Figure 5.18. *Main challenges for deploying the remote support tool were on data exploitation and on data traceability linked to end-to-end digital continuity*

5.4.3. *Virtual reality for inspection*

To carry out an inspection task, inspectors traditionally receive a quality check work order which contains instructions grouped into tasks and steps. These instructions very often refer to product documentation (drawings, digital mockups, specifications, etc.), which must be prepared and made accessible prior to starting the quality check. During the execution of the quality check, checklists guide the inspector through the product to guarantee certain key parameters (e.g. dimensions, final treatments, cosmetic appearance, test values, etc.), which are traditionally recorded on paper (or in isolated Excel spread sheets in the best cases) and need to be transcribed to create a specific report, or uploaded by populating specific work order fields, which doubles the workload without adding any specific value to the process.

Preparing product documentation, accessing it at the right moment and transcription of the collected real data into the official repository, are non-value-added tasks that can be improved. During proper execution of the quality check, inspectors usually have to manage paper documentation, drawings and lists that are sometimes difficult to manipulate in narrow spaces or areas with poor ergonomics.

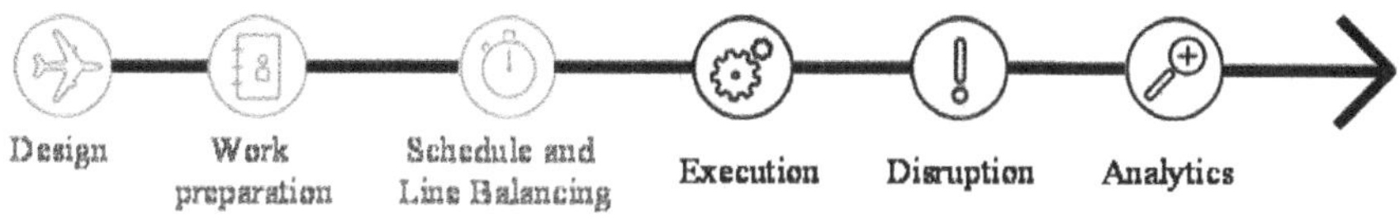

Figure 5.19. *Virtual reality for inspection of managed data at execution and disruption and will be a source of data in the near future for analytics activities*

Enriching the digital mockup with standard inspection criteria and creating a digital inspection work order, linked to the manufacturing execution system (MES) and enterprise resource planning (ERP) made it possible to give access to the right information at the right time; to guarantee that this access was achievable from the shop floor, hardware devices (tablets, glasses, tactile screens, etc.) were prepared and deployed into an industrial environment – which was not evident at all.

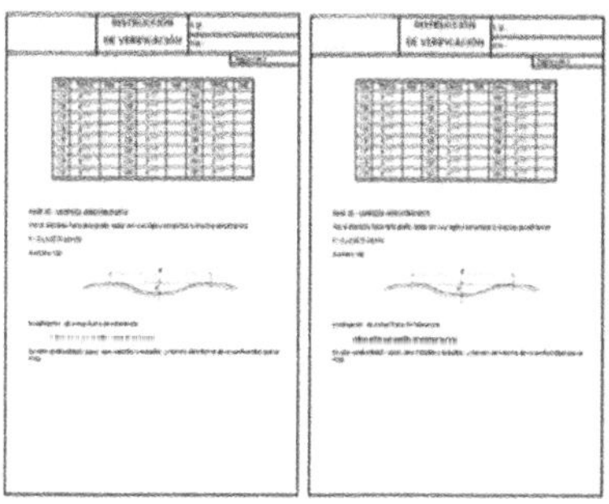

Figure 5.20. *Paper based verification instruction (left) and virtual reality exploitation for checking (right)*

This has resulted in an improved quality inspection process with easier visual comparison of physical product against digital mockup or drawings (using virtual reality and augmented reality), quick inspected part localization in digital mockups or product structure, immediate access to administrative process data (work order ID, step ID, manufacturer serial number, process step date, etc.), easy diagnosis by comparing virtual and real parts inspection representations, direct non-conformity creation during the inspection process and, finally, automation of the inspection by automatic report generation and automatic documentation upload.

During the implementation of virtual reality for inspection, geo-localization and image recognition techniques were the main challenges faced, in order to promote the self-orientation capabilities of the portable industrial tablets used.

Figure 5.21. *Main challenges for deploying virtual reality for inspection tool were on data exploitation techniques*

5.4.4. *Augmented reality (AR) for harness manufacturing*

Harness manufacturing in the aircraft and spacecraft industry is a highly customized, lead-time-focused activity (systems are very often driven by customers and final configuration is often decided late in the industrialization cycle). Harnesses and connectors are usually defined at the drawing level in a list of connections between dozens of wire pin terminals and their respective connector pin insert holes for plugging into. The traditional way of working consists of manual assembly of connections by consulting drawings, by looking for pin terminal and connector hole identification and by physically creating the soldered connection. If an incorrect connection occurs, it is not detected until the check phase at the end of the manufacturing process.

As a result of this manual labor, plenty of non-value added steps and connector manufacture were clearly identified as areas in which digitalization could definitely help. The aim was to develop an assisted visual system for the insertion of pins in to the connectors, using augmented reality for quick identification of the terminal to which the wire must be connected. To maintain data continuity, it uses real-time data exchange with the actual manufacturing execution system (MES) and enterprise resource planning (ERP). For the reduction of non-quality costs, due to insertion mistakes, image recognition techniques are used.

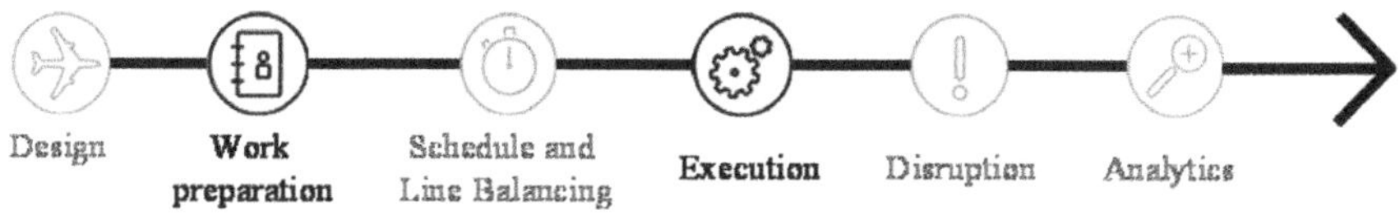

Figure 5.22. *From a data point of view, augmented reality for harness manufacturing mainly affects work preparation and execution activities*

From a technological perspective, the solution consists of a tooling fixture for locating QR codes and the connector to be assembled, and a camera connected to an industrial tablet, properly mastered and linked to the MES and ERP systems. To adapt the performance of the devices to the task, displays were added to visualize data corresponding to the work order and task being performed. In addition, transparency and zooming techniques were used in the image capturing and treatment to facilitate visualization and localization of small connector holes. The main challenges faced during the implementation were the correct visualization of information and feedback display and the real-time data exchange requirements.

Figure 5.23. *Augmented reality for harness manufacturing faced challenges on data exploitation and end-to-end digital continuity*

5.4.5. *Augmented reality (AR) for assembly*

For conducting harnesses and pipes along airframe and spacecraft structures, support bracket installation is the usual solution applied widely by design engineering offices. Those support brackets generally consist of a bent sheet metal or machined fitting with two differentiated parts: one used for supporting the pipe or harness and the other for fixing them to the aircraft or spacecraft structure. Fixing is usually done by using rivets or bolts, and as a general rule, the order of magnitude on the accuracy for the spatial positioning inside the structure is about 1 mm.

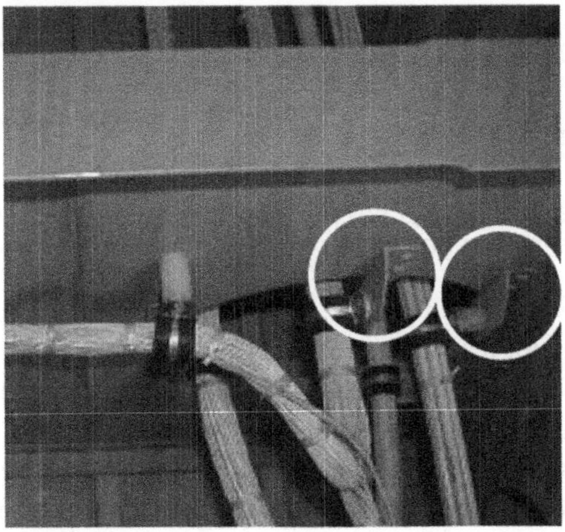

Figure 5.24. *Support brackets highlighted in white circles, conducting harnesses through a machined frame. For a color version of this figure, see www.iste.co.uk/duvaut/IAM.zip*

Usually the positioning of these supports is done by measuring distances from relevant structure items like frames or stringers, located in the surroundings, using traditional hand tools like rulers or laser markers. The distances for locating the support are usually stated in the documentation embedded into the work instruction. In the best cases, those data are directly available in the work instruction, but in the worst cases, those data are specified in drawings given in the work order, with the consequent effort of searching from one document to another to finally obtain the right value for either assembly or verification operations.

As locating and fixing these support brackets in fuselage assembly is one of the most time-consuming tasks in that kind of airframe (there are hundreds of them along stringers, panels and frames), clear benefits were identified in terms of non-value added task removal (and its consequent cost reduction or savings). Eliminating the document-searching effort was the first target, and manual measuring, marking and locating activities was the second. Furthermore, a third target was the removal of activities which involved repeating measurements, i.e. non-value added tasks.

The solution consisted of the implementation of a laser projection device, based on virtual templates, downloaded directly from the digital mockup and prepared subsequently. This device projects the profile and the rivet fixation position on top of the real aircraft surface. This allows the worker to perform the drilling operation and positioning of the support bracket directly on the structure, without needing manual measurements. The fact that the information to be projected comes from digital mockups frees the blue collar from the tedious task of navigating through work order documentation or, even worse, through drawings and sketches. Embedding this application into the digital work order eliminates the majority of documentation searching efforts.

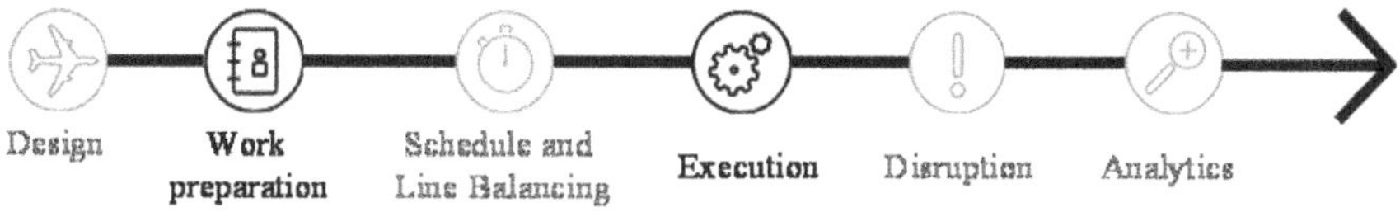

Figure 5.25. *From a data point of view, augmented reality for assembly mainly affects work preparation and execution activities*

Benefits of this implementation were clear from the perspective of a blue collar and inspector, as well as from the manufacturing engineering perspective. It represents direct information from DMU at shop floor level, directly shown over the aircraft structure, allowing hands-free support to blue collars and inspectors. Savings came from assembly times – a reduction in part positioning and non-quality cost reduction – there was a drastic reduction in the number of non-conformities due to incorrect part positioning. For inspectors, savings came from a reduction in documentation search times and a reduction in the time taken to check operations. For manufacturing engineers, even if documentation preparation time is longer initially, bigger savings came mainly from a reduction in assistance given to the shop floor. Reductions were also seen in the time taken to check configuration control and for non-conformance dispositions, leading to secondary savings.

From a technical point of view, the system is composed of both software and hardware. The software system consists of a programming module where the digital information to be shown is prepared; a simulation module, which avoids problems during the execution of the process; and a control module where the interface with the operator is built. The hardware system consists of a laser projection device, specific tooling for device location and manipulation, and an industrial tablet for launching the application interface with the blue collar or inspector.

When developing the solution, the main technological challenges were driven by visualization of information techniques. Also challenging was the automation of device localization, in terms of manufacturing serial number, assembly station identification, airframe position and stage in the work order identification. Other topics related to the proper mixed reality technology (graphic control, profile projection deformation control, calibration and prediction of shadows, etc.) were faced on the development of the tool. Widespread deployment of the solution challenged the capability for interfacing with existing IT ecosystems and the ergonomics for shop floor manipulation and usage.

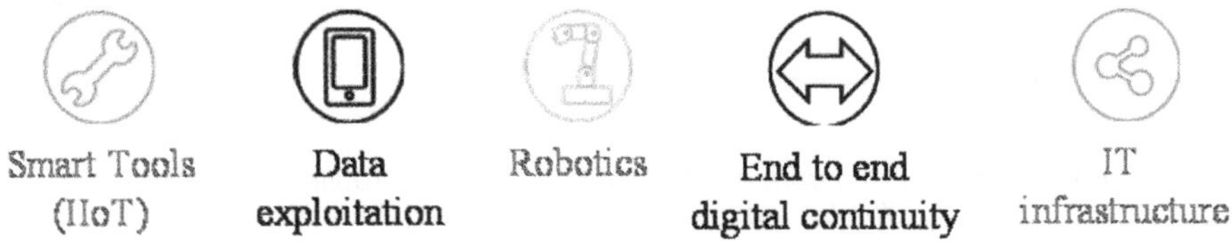

Figure 5.26. *Augmented reality for assembly faced challenges on data exploitation and end-to-end digital continuity*

The implementation of these digital initiatives has been the first step towards a digital transformation of the operations area inside Airbus DS.

Almost 2,000 digital SQCDP boards deployed, thousands of work orders digitalized into production, an IT architecture, four times faster, for massive connection to 3D data storage directly from the shop floors, and more than one thousand industrial tablets, touch screens and hardware devices delivered, are some figures that support this statement.

From an organizational angle, this deployment represents the synchronization of more than 15 plants in 18 months, with a harmonized way of working, focused on core technologies and business and under the umbrella of highly engaged top management.

However, the main transformation occurred in people's mindset: from an initial reluctance, small, quick-wins were engaging areas and staff. These early little successes empowered team members, consolidating and accelerating the process while facing new challenges and ups and downs the overall project. In-depth knowledge of the organization involved in the processes to be transformed was key to apply digitalization in a pragmatic and consistent way.

In parallel, research and development is being conducted to guarantee a long-term roadmap to fuel this transformation journey with new features, new capabilities and benchmark solutions. Future use cases will come when maturity reaches the right level, by introducing concepts like model-based design for aircraft systems, virtual reality for training purposes and virtual assistance to workers, based on chatbots and automated troubleshooting analyzers. We thank our past for all the lessons and knowledge acquired; we face the future innovating with a purpose.

Box 5.1. *Takeways*

5.5. Glossary and acronyms

AR: Augmented Reality

BOM: Bill of Materials

BSC: Balanced Scorecard

DMU: Digital Mockup

DWO: Digital Work Order

E2E: End-to-end

ERP: Enterprise Resource Planning

IIoT: Industrial Internet of Things

IT: Information Technology

KPI: Key Performance Indicators

LT: Lead Time

MES: Manufacturing Execution System

MRO: Maintenance, Repair and Overhaul

MSN: Manufacturer Serial Number

OES: Operational Excellence System

OLAP: Online Analytical Processing

PC: Personal Computer

PLC: Programmable Logic Controller

PLM: Product Lifecycle Management

SQCDP: Safety, Quality, Cost, Delivery and People

URL: Uniform Resource Locator

VR: Virtual Reality

6

The "AugmentChain": Sustainable Augmentation Value Chain for Intangible Assets

"The advance of technology is based on making it fit in so that you don't really even notice it, so it's part of everyday life."
Bill Gates

6.1. Are intangible assets the 21st Century's key value drivers?

Intangible assets (ITA) are the raw material for trade and social relationships in the 21st Century and, more precisely, for the value chains of the digital age (4.0). A product has become a "smart container", activated by software, which is the true source of its multi-faceted value: financial, societal, industrial, medical, well-being, cognitive, etc.

The following list mentions some ITAs that illustrate the extent of their penetration into all aspects of our lives:

– User experience (UX) and patient experience (PX), shaped by the digital traces of our pieces of connected life, which is now almost continuous in space and time and monetized by third parties, too often without our knowledge and without any real benefit to the user, apart from the promise of a hyper-personalized service:

- at home (while using Alexa or any connected home appliances), at work (through all the evaluations/screenings we undergo) and as a patient (in this case we refer to the patient experience, PX);

- in public transport, carpooling (BlaBlaCar), in your connected car, on scooters, bicycles, etc.;

- during our leisure activities and/or planning for them (tourism, games, social networks, etc.);

- looking for and using public and private services (post office, banks, insurance, mobile telephony, courier services, energy and water distribution, etc.).

– Reputation or quality of the customer relationship:

- evaluations or ratings – Airbnb, Uber, BlaBlaCar, Netflix, Amazon and Booking would have much less value without user ratings and reviews;

- word of mouth or recommendations allow many institutions to increase and retain their customers (Instagram/Facebook posts, YouTube videos, etc.).

– Knowledge, expertise, know-how, well-being, talent:

- the knowledge of a researcher, the expertise of an engineer, the know-how of a craftsman or a doctor/surgeon, the skill of a mediator or the talent of an athlete become intangible assets when they provide concrete services to society, industry, the health system, citizens, etc.

– Content of "collaborative innovation", powered by extensive (cross-company) ecosystems operating in real time and on short cycles of experimentation or implementation, involving four broad categories of stakeholders, namely suppliers, investors, public/private purchasers and users:

- ideas, architectures and presentations;

- problem formulation and specification;

- business models, human capital and organizational processes, etc.;

- proofs of income, feasibility, concepts and pre-products;

- trade secrets (law of July 30, 2018 in France).

– Contextualized data:

- in general, the core of ITAs, whatever form they take, consists of data, in sequences of 1s and 0s today, and Qbits tomorrow, with quantum computing, and "contextualized" – meaning scripted and associated with specific uses or known business parameters, such as mobility, sports, health, work, etc. – situations.

6.2. What are intangible asset value locks?

The creation, protection, exchange, valorization, enrichment and monetization of intangible assets all come up against obstacles in the form of three categories of systemic asymmetries of the digital economy and society between the sources of ITAs (citizens, patients, machines, etc.) and their third-party buyers/operators:

– *Trust and security asymmetries*:

- intrusion into our private life (privacy);

- weakness of authentication;

- unsuitable protection devices for 4.0 assets;

- absence of traceability of content and exchanges (e.g. when treating and tracking Covid-19 positive patients);

- lack of neutrality;

- during recent international presidential campaigns, a great deal of "fake news" has been found on social networks as well as in traditional newspapers.

– *Information and control asymmetries*:

- opacity: most digital platforms hide the procedure for terminating a subscription and deleting data;

- complexity: the terms and conditions imposed by technology companies require knowledge of the law to understand the implications of agreements.

– *Revenue and recognition asymmetries*:

- a platform such as Netflix is what we call an "exponential" business, i.e. one where turnover per employee exceeds $1 million ($4.5 million of revenue per employee in 2018, to be precise). In June 2019, for example, there were no exponential companies in France. Despite its huge size with 120,000 employees worldwide, Apple's revenue per employee was $2 million, which also placed it in the category of exponential companies. Netflix does not work miracles; however, it does exploit revenue asymmetries with a minimum number of expert employees who extract billions of hours' worth of video consumption to sell us an "on-demand service" aligned with our audio-visual tastes, without "rewarding" them. Without user experience, its marketing angle would be a blind spot;

- the ineligibility for the patenting of 80% of the innovation contents of the digital age greatly weakens the 4.0 innovators. For example, several 4.0 ITAs exist such as:

- business models,

- shared contributions in open innovation working, among communities and ecosystems,

- "anything as a service" concept, such as "mobility as a service", etc.

6.3. Can blockchain unleash the full value of intangible assets?

6.3.1. *Blockchain: the "trust machine" to neutralize human corruptibility*

It is no accident that the 2007–2008 financial crisis and the founding work of the "Bitcoin" blockchain by Satoshi Nakamoto – author of the first secure electronic peer-to-peer payment protocol, published on October 31, 2008 – were broadly simultaneous events.

As Joseph E. Stiglitz – Nobel Prize winner in economics – writes in his book *Freefall* [STI 10]:

> The crisis (of 2008) is not a disaster that would have happened to the financial markets anyway; it is man-made: Wall Street inflicted it on itself and the rest of society.

In response to what was a form of incompetence, or even a betrayal of the trust placed in individuals in charge of the economy and market supervision, a group of geeks developed a protocol delegating trust in the processes of cryptocurrency transactions (e.g. Bitcoins) to machines, which are freed from centralized and corruptible third-party creditors.

We have gone from one extreme to another in the world of transactions, from "entirely human" to "entirely machine": a circle of institutions under the control of opaque and incomprehensible financial instruments – for example securitization of subprimes – to distributed computerized registers of securities exchanges in the broad sense (currencies, assets, intangible assets, etc.), tamper-proof, resilient, traceable and based on the disintermediation of banks and historical monetary instruments (extracts from "Understanding the Block Chains", report of the OPECST (*Office parlementaire d'évaluation des choix scientifiques et technologiques* – Parliamentary Office for Scientific and Technological Choices), June 2018).

Ten years after the advent of Bitcoin (BTC), specializing in money transactions ($3 trillion in volume traded in 2018 or 3.5% of the gross world product, estimated at $88 trillion the same year), there are more than 1,500 cryptocurrencies (e.g. ETHER associated with Ethereum), and applications of the "Trust machine" – as *The Economist* called it – abound, in areas such as energy (LO3 Energy), banking and insurance (R3-Corda), logistics (ADEPT), health (see e-Estonia for healthcare), etc.

The 10th anniversary of Bitcoin also marked the launch of the first blockchain "Marketplace" at the CES in Las Vegas, held on January 8–11, 2019, a sign that practical solutions are emerging for users and consumers alike. Ginny Rometty, CEO of IBM, and Charles Redfield, Executive VP Food at Walmart, presented "IBM Food Trust" (based on Hyperledger), a true chain of trust for food traceability, bringing it to the attention of 265 million Americans. With the scandal of spoiled meat imported from Poland[1] happening at the same time, consumers were sure to appreciate it.

6.3.2. *A hybrid human–machine trust protocol*

Like human–machine cooperation (or co-evolution) catalyzed by the virtualization and cognitive engines of digital transformation (data, IA, IoT, 5G, bots, RV-RA, digital twin, etc.), as seen in augmented medicine, for instance, the understanding of the issues inherent in transaction validation protocols is based on their reading in terms of the degree of the machine's involvement in the "delegation of trust".

---

1 *Les Échos*, February 1, 2019.

The historical benchmark, before blockchain, is based on "all human" or 0% machine in the delegation of trust. Bitcoin has set the so-called "public" protocol as standard, which we will consider, out of convention, as 100% machine, so anyone can write, read or audit the transaction's computer records. In the so-called "private" protocol, the members of a consortium decide who reads, writes and audits the registers distributed, as well as the related operating arrangements; this is used, for example, in banking and insurance. This is a true "hybrid human–machine trust protocol". Blockchain purists, in line with the Cypherpunks of the late 20th Century, supporters of a global network of fully-open anonymized and encrypted transactions, also challenge the arbitrary and centralized involvement of private blockchain consortia.

The question of the degree of human–machine hybridization of a protocol of trust is not only political or philosophical. A 100% machine version is very expensive. The original Bitcoin protocol, called proof of work (PoW), aims to prevent malicious people from attacking the Bitcoin network by forcing them to use their computing power to prove their contributions to the network. PoW relies on "miners" who validate and secure transactions through computing power.

The original "proof of work" (PoW) suffers from eight major systemic limitations:

– carbon footprint/energy consumption. According to Ronan Le Gleu – French Senator and co-author of the Blockchain report commissioned by OPECST, June 2018 – "we estimate that Bitcoin alone consumes at least 45 TWh per year, equivalent to 8.5% of France's energy consumption in 2017, estimated at 530 TWh";

– complexity of ownership/usability both for end-users and for decentralized applications developers. A Bitcoin interface is not as easy to handle as contactless payment, and it really needs to be, at least for the sake of citizen empowerment;

– slowness, as PoW needs an average of 10 minutes to validate a transaction;

– scalability, as PoW validates just a few transactions per second, which is incompatible with the real-time needs of trading and 4.0 exchanges;

– cost of transaction/validation on Bitcoin is tens of US dollars. In addition, this cost is the same whether the transfer is for $1 million or $50;

– fee volatility due to network congestion slows down the validation process; associated fees may vary on Bitcoin by up to a factor of 10;

– privacy, as public transaction validation protocols (such as PoW) cannot, by design, maintain complete privacy for relevant information;

– the lack of a legal basis for the "smart contacts" which, according to Marco Iansiti, the David Sarnoff Professor of Business Administration at Harvard Business School, "automate payments and transfer of currencies or other assets, once conditions are negotiated". In the event of litigation surrounding the performance of a smart contract, none of the parties can avail themselves of a legally-enforceable guarantee.

6.3.3. *EOS, a path to solve some systemic blockchain issues*

On June 14, 2018, Daniel Larimer, originally from the community network Steemit, launched with great fanfare – and an ICO[2] worth $4 billion according to the Wall Street Journal – the EOS blockchain that aims to supplant Ethereum by solving the lack of systematic scalability with a spectacular performance: EOS claims 300,000 transactions per second, compared with seven for Ethereum. EOS replaces the energy-intensive PoW by the delegated proof of stake (DPoS) that mandates the "human" part of a hybrid trust protocol to the "crypto-shareholders" of EOS, holders of "EOS tokens" or digital assets in EOS cryptocurrency.

Many other distributed trust protocols have been undertaken to solve PoW limitations, and it is outside the scope of this book to go through them thoroughly. Interested readers can refer to the IXXO white paper for more details[3].

On September 12, 2018, article 26 on ICOs – and thus on token issues – was adopted under the PACT law passed in France. Bruno Lemaire described it as follows:

2 ICO (Initial Coin Offering): a cryptocurrency fundraiser that allows blockchain startups to gain quick access to financing.

3 https://ixxo.io/IXXO_White_Paper_2_2_SMALL.pdf.

A legal framework for ICO is created. The *Autorité des Marchés Financiers* (AMF) will be able to issue a visa to players respecting the protection criteria of savers. This legal framework will attract innovators from around the world.

6.3.4. *In spite of massive R&D worldwide and an explosion in use-cases, blockchain is still – after 10 years of existence – Tech4Geeks4Speculation!*

It is clear that, at first glance, a decade of existence and innovation has not succeeded in changing blockchain from a technology for geeks and speculators into a consumer service, centered on the needs of people and users, one as simple, ergonomic and cheap as online payment devices or the provision of multimedia content on open platforms.

This is a huge paradox, for two reasons!

First, blockchain technology is considered "just as disruptive as the TCP/IP protocol that led to the launch of the World Wide Web in the 1990s" (*Harvard Business Review*, "The Must of the Digital", Spring 2018) with the additional capacity, over the Internet, to give people and users a form of power and control ("empowerment") over economic, political, societal and legal practices through the hybridization of trust.

Second, there has been a tremendous demand from users for a secure Internet of Trust since the pivotal year 2019. It is exactly at this turning point that cognitive colonization was revealed (see, for instance, *The Attention Merchants* by Tim Wu).

6.3.5. *Towards GreenTouch4Good and a sustainable digital: the "Internet of Augmented Me, I.AM"!*

The first 30 years of the digital era, since the World Wide Web officially launched in 1990, might be viewed as the era of "Datystopia" (data + dystopia). Americans check their smartphones more than 2,500 times a day and the American population clicks "like" on Facebook 4 million times per minute. Citizen-users are demanding sustainable and responsible digital tech that gives them not only control over the use of their data but also over their intangible assets (which is the intention behind the GDPR[4]). They need the appropriate tools to protect value and possibly monetize their ITA – this would be the role of a new generation of blockchains. The economics of the major service platforms, such as GAFA (Google, Amazon, Facebook, Apple) and NATU (Netflix, Airbnb, Tesla, Uber) in the USA and BATX (Baidu, Alibaba, Tencent, Xiaomi) in China, conferred a status of true "intangible asset" on the UX (User Experience), e.g. photos, videos, conversations, impressions, written content. This is now established as raw material for trade and social relationships in the 21st Century!

6.3.6. *Emergence of "empowerment driven" blockchain protocols and usage*

The year 2019 – whether at CES in Las Vegas or MWC in Barcelona – marked the arrival of new trust protocols that attack the historical limitations of blockchain by democratizing its uses, centered on citizen empowerment. For example, Pundi X presented its XPhone, which makes it possible, in blockchain mode, to surf the Net, exchange SMS and, in short, have a fully decentralized mobile life within the reach of a touch screen. Browser Brave and Mercedes-Benz reward virtuous behaviors – on the Net or while driving – with tokens. Uniris has launched a co-responsible blockchain. IMT (the French equivalent of MIT), in partnership with the startup IXXO, has launched a "JuriGreen" blockchain at the end of 2019, based on the mixed proof of authority (MPoA[5]) protocol that eliminates all major blockchain issues with its small carbon footprint, agility and inclusivity, including a form of legal basis, towards sustainability and the Internet of Augmented Me, I.AM.

JuriGreen Blockchain, combined with the MPoA protocol, appears as a key enabler for safe and ethical Covid-19 containment, exit or partial containment, allowing the anonymous tracking of positive patients while

4 General Data Protection Regulation, enforced by the EU in May 2018 for all data collected from EU citizens.

5 PoA: Proof of Authority. In PoA-based networks, transactions and blocks are validated by approved accounts, known as validators. Validators run software to allow them to put transactions in blocks.

MPoA: Mixed Proof of Authority. MPoA allows the user (application or single user) to choose the cluster associated with the required level of trust and transaction security. The cluster fees depend on the provided trust and security level. "Mixed" clusters are put in competition to increase "decentralization" virtue.

complying with full privacy, in keeping with the GDPR and regional health regulations, controlled by the patients!

6.3.7. *Blockchain map*

Figure 6.1 shows a summary view of a blockchain map from Tech4Geeks4Speculation (top right quadrant), which started in 2008, towards GreenTouch4Good (bottom left quadrant), which can contribute to a "sustainable" intangible asset value chain, for the coming decade, the 2020s. The map puts together most of the information provided in this section.

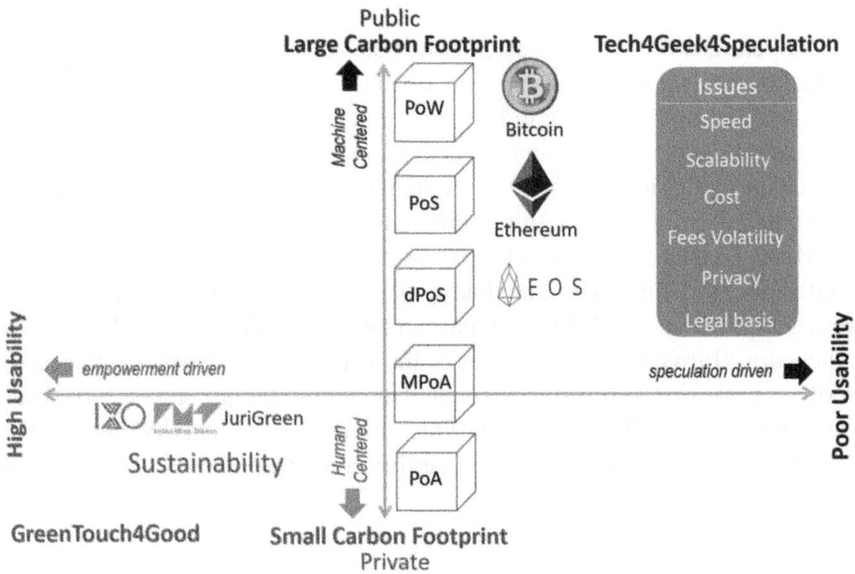

Figure 6.1. *Blockchain map*

The vertical axis represents the first of the eight major issues to be tackled when dealing with the carbon foot print/energy consumption. Public protocols, such as proof of work (PoW), or proof of stake (PoS), and the delegate proof of stake (DPoS), are more "machine centered" and consequently burn large amounts of energy. Private protocols, such as proof of authority (PoA), are more "human centered" and can achieve a much lower carbon footprint. The mixed proof of authority (MPoA) probably represents the best of the hybrid "human–machine" validation protocols with an adjustable/smart degree of human involvement according to the underlying value of the transactions and the amount of energy that can be used.

The horizontal axis represents the usability issue. Most speculation driven use cases, such as Bitcoin, Ethereum and EOS, are handled by geeks who are not bothered about poor usability (see the right-hand side of the figure), whether for end-users or decentralized application developers. Empowerment driven use cases that are expected to boost horizontal governance and citizens' rights, such as early and low-cost ITA protection, up to a click by their owners, are poised to bear very high usability "by design". The other six major issues to overcome, towards a GreenTouch4Good, which was explained earlier in this section, are listed in the top right quadrant.

While "empowerment driven" blockchains – such as JuriGreen, launched by IXXO and IMT, with its small carbon footprint and the fact that it also eliminates all the other limitations of speed, scalability, cost, fee volatility, privacy and legal basis (see the bottom left quadrant in Figure 6.1) – can contribute to unleashing ITA's multi-faceted value (financial, societal, cognitive, medical, industrial, environmental, etc.), they cannot do so alone. They are only one of many bricks needed to build the whole sustainable ITA value chain.

6.4. What is the intangible asset AugmentChain?

6.4.1. *Continuum of three worlds: physical, digital and biological*

Whether you drive the latest connected car U5 from AIWAYS[6] equipped with a genuine cockpit for the best in-vehicle experience, or whether you wear the smart Chronolife T-shirt[7] to continuously monitor your critical health parameters, you will experience a continuum of three worlds (physical, digital and biological), interacting seamlessly together in real time to anticipate and deliver the most valuable ITAs: your next need or wish, even if you are not yet aware of it!

6 https://www.ai-ways.com/eng/u5.html.
7 http://www.chronolife.net/monitoring-solution.

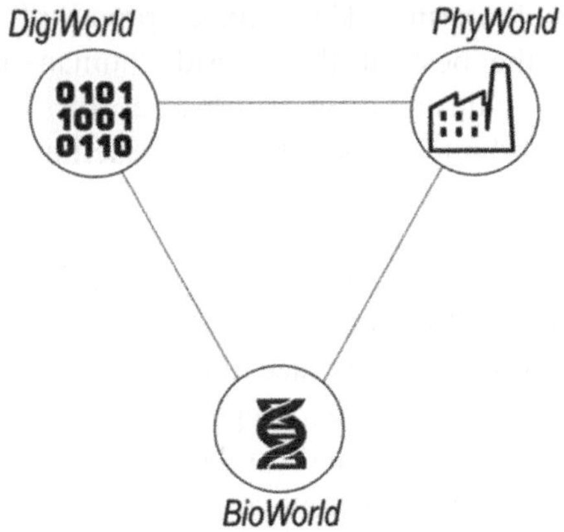

Figure 6.2. *"Augmentation age": real-time interaction and co-evolution of the three worlds into a "BioPhyGital" world*

In the case of the U5 SUV, the physical world comprises all the parts of the car, i.e. the engine, wheels, seats, metal frame, composite materials, glass, rubber, wood, etc. In addition to the billions of bytes of data collected per second in the U5 ecosystem, including all its connected parts, other neighboring vehicles and infrastructure networks involved in safe driving and entertainment services, the U5 digital world includes all the software code converting all the car's parts into smart objects, which become true "living and secure assets". All U5 passengers and the driver are the inhabitants of the biological world.

As regards the Chronolife product, the T-shirt itself spans the physical world along with its embedded physiological sensors. The associated digital sphere contains (similar to the U5 case) software and data, which are collected both locally and remotely. The person wearing the T-shirt is the biological world. In some similar use cases, the biological world can be living cells, plants, animals, organs or the brain itself, such as Neuralink[8].

Obviously, based on the above two descriptions, and paradoxically, because of their obviously different purposes, the U5 and the smart T-shirt share many common features, and the same would hold true for the tens of thousands of services/UX brought by trillions of connected objects

8 https://www.neuralink.com/.

worldwide, activated by billions of lines of software, in any facet of our life, at home, at work, on the move, consuming entertainment/sport, at the doctor's or hospital, and for any other kind of activity you are engaged in.

> Thanks to powerful digital engines, we are now engaged in a co-evolution or convergence of the biological, physical and digital worlds into a "BioPhyGital" world (see Figure 6.2) that is permanently connected, striving to deliver augmented UX in the form of existential intangible assets, such as your next need or wish, which we can refer to as the Internet of Augmented Me, I.AM.

How can it be done? Is there any underlying process that drives these three worlds together in an efficient, responsible and resilient way? One answer is the augmentation convergence of these three worlds based on the AugmentChain, i.e. the sustainable augmentation value chain for intangible assets.

6.4.2. *The AugmentChain is propelled by three "pyramidal" engines*

As shown in Figure 6.3, the augmentation convergence of the Bio, Phy and Digital worlds is based on the AugmentChain propelled by three "pyramidal" engines, the enablers, the catalysts and the outcome that work together in short cycles (typically a few months). The AugmentChain is going through enrichment loops of experimentation (more precisely "effectuation") that leverage feedback from each of the three worlds (Bio, Phy, Digital) to continuously improve the efficiency and the resilience of the user experience delivered.

As will be detailed further on, the three propelling engines are described as "pyramidal" since each of them is fueled by four complementary sources.

> Effectuation was designed in 2001 by a cognitive scientist named Saras Sarasvathy. Originally aimed at entrepreneurs, as a kind of "reverse logic" way of thinking, effectuation provides a way to control a future that is inherently unpredictable. In brief, instead of setting goals first and then looking for the means to achieve them, which is known as causal logic, effectual logic starts with what you have in your hands and then sets goals according to what you can afford to lose.

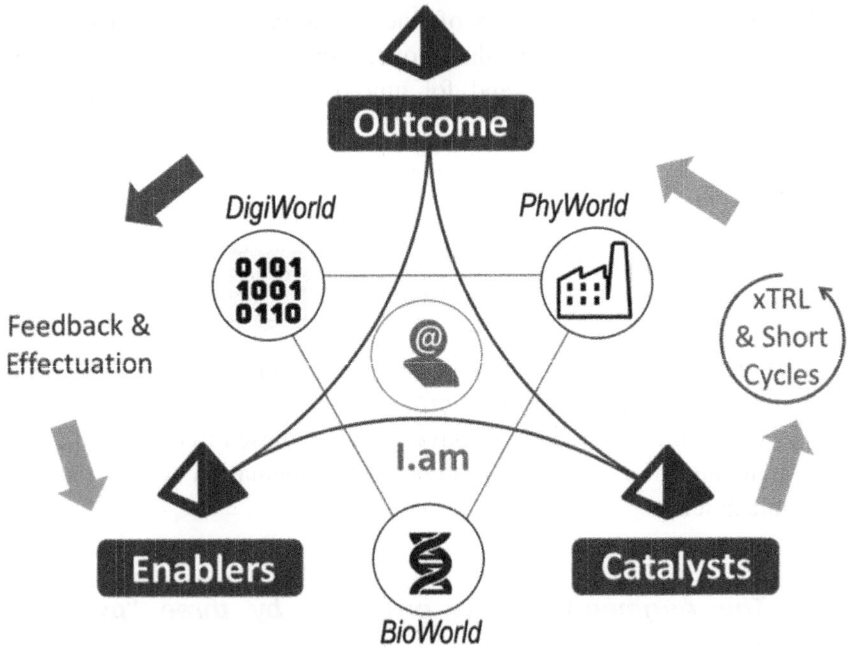

Figure 6.3. *The augmentation convergence: three "pyramidal" engines that propel the AugmentChain towards the Internet of Augmented Me, I.AM (xTRL – Multiple Technology Readiness[9] Level)*

This efficient and resilient effectual logic can be applied to any creation process and serves as a key principle of the AugmentChain (see above)[10].

6.4.3. *The outcome pyramidal engine*

As Gary Shapiro, CEO of CTA (Consumer Technology Association), pointed out in his opening talk in Las Vegas at CES 2020 (the world's biggest tech gathering worldwide), "Every company must now embrace technology to succeed."

9 Technology Readiness Level gauges the maturity of technologies during the acquisition phase of a program. It was developed at NASA during the moon program in the 1960s–1970s. For more details, see https://en.wikipedia.org/wiki/Technology_readiness_level.

10 For more information, see https://www.effectuation.org/sites/default/files/documents/effectuation-3-pager.pdf.

But beyond technologies, even transformational technologies such as AI, blockchain and quantum computing[11], the foremost expectations for humans, companies, organizations, society, the economy and, more generally, for the whole planet, are the outcomes (in terms of multi-faceted value) reshaping and disrupting our lives, as a result of the co-evolution of the three worlds, i.e. Bio, Phy and Digital.

Figure 6.4 shows the four fueling components of the outcome engine that we have decided to highlight. We understand that the choice of these four elements may be quite arbitrary and short-lived, in spite of the more robust AugmentChain generic and modular model shown in Figure 6.3. Several readers would probably choose different practical realizations. Incidentally, based on the generic model, these readers could build and operate their own AugmentChain!

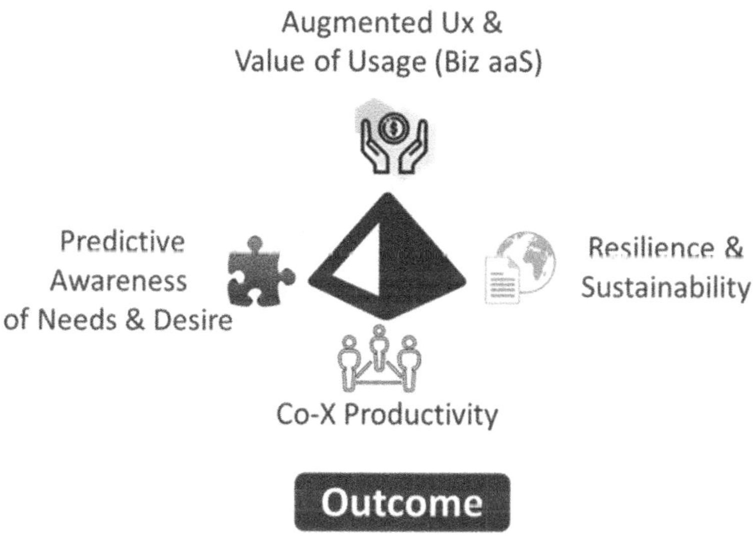

Figure 6.4. *The AugmentChain outcome pyramidal engine*

In particular, we have purposely skipped basic and trivial value-driven outcomes such as cost reduction, increased revenues, improved efficiency and new business models.

11 Quantum computing chips use quantum mechanics phenomena to achieve calculation performance that surpasses the most advanced current neurochips, by several orders of magnitude.

Motivated by the quest for a new sustainable digital era where major platform asymmetries (lack of trust, information and benefits, as sometimes encountered in GAFA[12], NATU[13] and BATX[14]) could be balanced out, the outcomes mentioned above are more driven by sustainability, sense, well-being, usage, humans, citizens' rights, etc.

Moreover, achieving all these particular goals via an ITA value chain would surely unleash the power of exponential revenues.

The four vertices of the outcome pyramidal engine are definitely very valuable intangible assets (shown in the clockwise direction in Figure 6.4):

– predictive awareness of needs and wishes: the idea is to be one step ahead. For example, Beyond Verbal[15] has developed a disruptive technology that enables emotions, well-being and health conditions that are conveyed through the human voice to be understood. "We envision a world in which personal devices understand our emotions and well-being, enabling us to become more in tune with ourselves", says Beyond Verbal CEO Yuval Mor;

– augmented user experience (UX) and value of usage (business as a service): what people want – whether they are at home, on the move, at leisure, at work, etc. – is the best possible user experience in the shortest time, at an affordable cost, even though, in the smartphone market, average customers are willing to pay $1,000 for a device that will become obsolete in a few months! For example, Moovel[16] provides mobility as a service by optimizing the process of selecting the most suitable means of transportation (including bikes, public and private transportation, eScooters, etc.) and paying for the trip with one click, while avoiding traffic jams;

– resilience and sustainability: the goal is to include sustainability and resilience by design in the ITA value chain for absolutely any vertical, in order to keep the world fed, secure, safe, healthy and powered, whether or not we deal with continuous and dramatic erosion of the planet resources and its pollution or natural disasters, such as tsunamis and earthquakes.

12 Google, Amazon, Facebook, Apple.

13 Netflix, Airbnb, Tesla, Uber.

14 Baidu, Alibaba, Tencent, Xiaomi.

15 http://beyondverbal.com/.

16 https://www.moovel.com/en.

For example, Zero Mass Water[17] uses solar-powered hydro panels to condense water from the air molecules;

– co-X productivity (X stands for any activity engaged jointly among a group of people, such as creativity, innovation, working, thinking, moving, caring, protecting, etc.): companies worldwide, such as PWC, Microsoft and American Airlines, have become more collaborative to accelerate their innovation. They make use of a number of co-innovation platforms such as Chaordix[18], Brightidea[19] and Spigit[20].

6.4.4. *The AugmentChain enabler's pyramidal engine*

The enabler's engine shown in Figure 6.5 is fueled by disruptive technologies that make the convergence of the three worlds into one "BioPhyGital" world possible, rewiring entirely the way we think, work, play, care, build, spend money, travel, make friends, meet our mates, date, find jobs, study, communicate, etc.

Powered by a tremendous push for ever higher capacity chips, as Moore's Law[21] is reaching its limits, thus requiring the invention of new neurochip[22] (today) and quantum-like[23] technologies (tomorrow), technologies reshape our lives and change every couple of months. The technologies shown in Figure 6.5 are based on CES 2020 trends. Of course, some of them might have become obsolete in a few years, replaced by stunning new breakthroughs that will surely come in the future even though they remain as yet unknown.

Nevertheless, beyond the short-lived technologies themselves, the "anthropomorphic" way the above technologies are organized in a pyramidal fashion holds true more permanently.

---

17 https://www.zeromasswater.com/select-your-location/.

18 https://www.chaordix.com/.

19 https://www.brightidea.com/.

20 https://www.spigit.com/.

21 Moore's Law is the observation that the number of transistors in a dense integrated circuit doubles about every two years.

22 See, for example, https://anotherbrain.ai/.

23 Quantum computing chips use quantum mechanics phenomena to achieve calculation performance that surpasses the most advanced current neurochips, by several orders of magnitude.

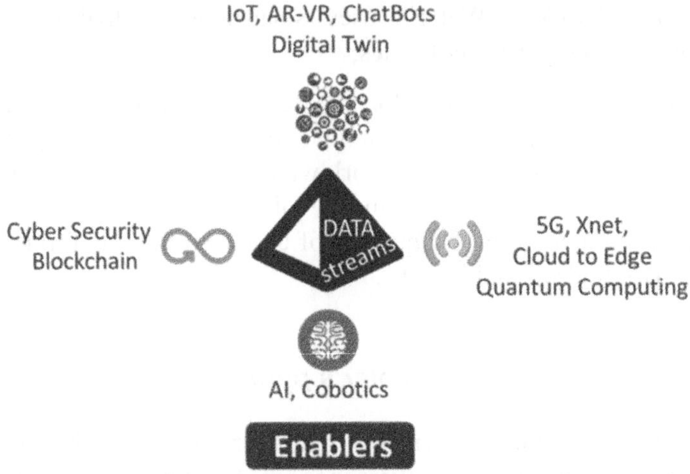

Figure 6.5. *The enabler's AugmentChain pyramidal engine*

In fact, each of the four vertices of the enabler's pyramid, shown in Figure 6.5, mimics key human capacities that have triggered and fostered the evolution of humankind for hundreds of thousands of years. It is therefore understandable that the sophistication and the performance of this mimesis will always evolve and improve, while the underlying basic human capacities that it imitates will most likely be forever part of our DNA, at least for quite some time. Moreover, even though they might change, they will do so much slower than these technologies.

Moreover, as with the other two AugmentChain engines (outcome and catalysts), the pyramid symbol materializes the fact that the components standing on the four vertices are – thanks to the data streams collected – integrated and connected together, always and in real time, and are continuously learning from each other towards global augmentation convergence.

We start from the top vertex and move clockwise. As it will turn out, below this order, there follows a number of sequences or routines that we unconsciously perform thousands of times every day.

IoT, augmented reality–virtual reality, mixed reality MR (AR + VR), chatbots and digital twins. These latest technologies mimic the crucial "**sensing and interacting**" capacities of humans that enable the perception of the world around us and the possibility of continuously interacting with it.

The IoT (several trillions of everyday objects connected to the Internet) is the largest network ever created (and is growing exponentially). It consists of smart devices that emulate our senses, eyes (eye-tracking digital cameras[24]), ears (3D smart sound capture devices[25]), smell and taste[26], touch (tactile interactions[27]) and beyond our senses, our feelings and emotions[28] and even beyond that, thousands of measured physical, chemical and biological quantified features of our environment[29].

The IoT may be viewed as the common gigantic sensory device and network of the "BioPhyGital" world. Similar to humans, who collect information from their environment and memory before undertaking any type of activity or triggering any kind of function, the IoT generates tons of data streams that serve as a connecting fuel of the enabler's pyramidal engine. Data is the greatest natural resource of the BioPhyGital world. Mark Liberman calculated the storage requirements for all human speech ever spoken at 42 zettabytes[30] if digitized at 16 kHz for 16-bit audio.

Augmented reality (AR), virtual reality (VR), mixed reality (MR), chatbots, earbuds and digital twins are probably the most powerful technologies in 2019, which, combined in a joint approach, unleash a fast, efficient, accurate and easy way to visualize and understand interaction in the BioPhyGital world.

Thanks to the IoT, and based on the collection of all data related to one particular target, whether it is a person, an object, a process, a phenomenon, etc., it is possible to build a digital double (or twin) of almost anything (machines in industry, patients in Medicare, vehicles in traffic management, customer behaviors in e-commerce, practice styles in sports, spending trends for bankers, etc.) that will act as a single source of adaptive truth for any task

24 http://www.tobii.com/.
25 https://visisonics.com/.
26 http://www.aromyx.com/.
27 https://www.ultrahaptics.com/.
28 https://www.affectiva.com/.
29 https://www.aeroqual.com/.
30 A zettabyte is 10^{21} bytes.

you have to undertake or any decision you have to make (see, for example, Visualiz)[31].

A digital twin is the generic and continuously updated representation of anything or anyone, enabling us to seamlessly interact and "see" in the BioPhyGital world.

Staying in the purely cyber sphere (digital), sometimes called the "Metaverse"[32], digital twins will be conveniently converted for visualization into virtual reality, accessible with special equipment such as a VR helmet (see Oculus)[33]. Augmented reality is an overlay to your own physical world in a 3D digital representation, leading to better understanding and anticipation. Augmedix, for example[34], provides physicians a hands-free way of pulling up patient data and displaying them on smart glasses, increasing facetime with patients.

In Industry 4.0, digital twins can be used at any step in the value chain. They can boost production efficiency by reducing the cost of a bricks-and-mortar building, built only after accurate design and tests in cyberspace. Digital twins also play a key role in predictive maintenance, reducing downtime and the cost of replacement, and boosting workers' productivity. Digital twin is applied by companies such as IBM, SAP, Siemens, Microsoft and GE.

Voice user interface (VUI) has fully penetrated our lives, such as Alexa, Siri and Google Home. VUI are everywhere, in all domains[35]: smart mirrors, connected cars, lighting, TV, speakers, alarm clocks, thermostats, headphones, faucets, AR glasses, etc. According to Lauren Foye, by 2022, banks will automate up to 90% of their customer interaction using chatbots. A survey conducted by Spiceworks showed that 40% of large companies employing more than 500 people plan to implement one or more intelligent assistant or AI-based chat robots over corporate mobile devices in 2019.

---

31 http://visualiz.io/.

32 Metaverse is a virtual world constructed on the Internet where people can interact freely in a variety of ways and forms. By using technologies such as AR (augmented reality), VR (virtual reality) and MR (mixed reality), Metaverse is able to connect the digital world with the real world.

33 https://www.oculus.com/.

34 https://www.augmedix.com/.

35 Source: CES 2019.

At CES 2020, Nissan unveiled an invisible-to-visible (I2V)[36] interface for its connected cars. I2V is another example of the augmentation convergence of the three worlds. It uses a 3D augmented reality interface that merges the real world and the virtual world to make information – which the driver would not otherwise see – visible. Information collected by sensors inside and outside the vehicle is combined with data from the omni-sensing cloud to provide the driver with enhanced information about the surrounding area, including predictive information ahead of the vehicle. This includes potential hazards hidden behind buildings, and obstacles ahead of a blind corner – all displayed seamlessly to the driver to support a confident driving experience.

One striking fact is that, since the advent of computers in the 1960s, human-to-machine interfaces have steadily become closer and closer to the body, blurring the boundaries between the Phy and Bio worlds. Indeed, we have in turn experienced punch cards, then keyboards, kinetic interactions, touchscreens, wearables, and now "bodyables" (brain to machine and body to machine)[37].

Data streams collected by the AugmentChain enabler's engine, shown in Figure 6.5, via its top vertex, either directly with IoT digitization or through numerous interactions, based on AR/VR, chatbots, earbuds and digital twins, are permanently undergoing another human-like operation, dealing with communication and information processing.

5G, X[38]-Net, cloud to edge, quantum computing. The emerging new 5G telecommunication protocol deployed for the first time in the world in Shanghai in March 2019, all social networks (X-Net, see footnote 15 of this chapter), the adjustable data processing between cloud and the edge of networks, and the upcoming quantum computing; all these technologies listed on the right vertex of the enabler's pyramid in Figure 6.5 mimic the special "**communicating and processing**" capabilities of humans. Beyond academic disputes about why, when and how human-like languages appear and evolve.

> The ability of men and women to communicate with one another in intelligent, symbolic, often abstract speech and

36 https://www.digitaltrends.com/cars/nissan-invisible-to-visible-tech-ces-2019/.

37 See Kernel: https://kernel.co/.

38 X stands for any type of network, whether it is a social network such as Facebook, Twitter, WeChat by Tencent in China, or professional networks. Of course, these social networks go along with all information exchange applications such as WhatsApp and Instagram.

> writing is the great unbridgeable gap between all humankind and every component of the animal kingdom.

As Henry M. Morris points out[39]. An interesting question is whether this fact will hold true for machines?

Indeed, 5G network and communication protocols first aim to enable the gigantic sensory devices and networks of the BioPhyGital world, i.e. the IoT, to be equipped with almost human-like ubiquitous and efficient abilities to seamlessly communicate, both within IoT networks and across IoT and human networks. In essence, 5G deals with "machine-to-machine" (Phy2Phy) and "machine-to-human" (Phy2Bio/Bio2Phy) communications, after 4G endeavored to bring the best performance it could for conventional human-to-human (Bio2Bio) communications. Since, as explained above, nowadays we interact best with the Phy world (Bio2Phy) via AR/VR/MR technologies, 5G faces the huge challenge of seamlessly supporting very high-speed wireless communications for real-time VR/AR/MR. As Steve Mollenkopf from Qualcomm puts it[40], "5G will take us from an online shopping experience to an immersive shopping experience."

Moreover, when connected autonomous vehicles check what is concealed on the road over the top of a hill using the I2V (invisible-to-visible) interface, the 5G response time, known as "latency", must be fast enough to warn the driver and vehicle that a car is out of control in the wrong lane and needs to be dodged to save lives (V2X[41] communications). The same holds true for decentralized and robot-assisted surgery where delay/latency sometimes means brain survival time. According to the ITU[42], the following KPI[43] should be achieved by 5G:

– speeds of 20 gigabits[44] per second for enhanced mobile broadband (eMBB). This very high speed enables many use cases such as AR/VR/MR everywhere, including eye-tracking technology[45];

– 1 million served users per square kilometer, for massive machine-type communications (mMTC) with applications such as I2oT[46] communications

39 https://www.icr.org/article/mystery-human-language.

40 Source: CES 2019.

41 Vehicle to anything, could be V2V, V2Infrastructure, etc.

42 International Telecommunications Union.

43 Key Performance Indicator.

44 1 gigabit is 10^9 bits.

45 See https://vr.tobii.com/products/htc-vive-pro-eye/.

for Industry 4.0 or the possibility to reliably and safely perform air traffic control for millions of drones[47];

– 1 ms latency for ultra-reliable low-latency communications (URLLC), typically for V2X networks and augmented surgery (see the 5G-enabled hospital in Chicago)[48].

Last but not least, just like humans adapt the way they communicate according to the situation and the interlocutor, 5G includes some smart capabilities that allow it to power the performance appropriate to the use case. We call this "usage-driven connectivity", as detailed in the three examples above.

Understandable and articulate communication cannot exist without some kind of "processing" of information to be transmitted and received. Our eyes can peer carefully at a moving crowd at the airport or railway station to eliminate background events and focus only on human faces. Our ears have the unique skill to naturally de-noise sounds when somebody talks to us at a concert or on the subway platform while a noisy train is just entering the station. In a permanently and fully connected BioPhyGital world, processing capabilities are based on powerful chips that have gradually migrated from remote cloud servers to the edge of networks, embedded in the connected/smart objects of the IoT, noticeably AR/VR/MR devices, drones, cameras, security cameras, vehicle parts, wearables, "bodyables", etc. By moving processing power closer to data sources, latency is dramatically reduced for improved user experience. The bandwidth needed is also reduced. Take security camera networks in a warehouse: instead of having to continuously transmit pictures to a cloud-only centralized surveillance center, edge processing will transmit pictures only when something is moving in the warehouse, since everything should be quiet if no breaches have occurred. The race is on among large companies, such as NVIDIA, Intel and Qualcomm, for the most powerful edge-enabling chips with the lowest power consumption possible. While preventing data from going to the cloud improves privacy and the compliancy with GDPR[49], it nevertheless raises numerous security issues[50].

46 Industry Internet of Things.

47 See, for example, https://www.airspacedrone.com/en/.

48 https://www.rush-health.com/RHAWeb/default.aspx.

49 GDPR, General Data Protection Regulation.

50 See https://www.journals.elsevier.com/computers-and-security/call-for-papers/special-issue-on-security-and-privacy-for-edge-computing-cu.

Gordon Moore, co-founder of Intel in 1968, devised Moore's Law[51] which states that the number of transistors the industry is able to place on a silicon chip doubles every 18–24 months. This guideline has governed the trajectory of the multi-trillion dollar semiconductor industry since then, for five decades. Moore's Law conversely means that the smallest feature of a transistor should decrease at the same rate! The digital era has been driven so far by coding and processing information on chips, in sequences of bits that can take only two values (1 or 0), with the constraint that each bit requires one transistor whose electrical state, acting as a switch, sets a status of either 1 or 0. According to Gordon Moore's predictions, by 2022, the smallest feature of a transistor should reach the size of a few atoms, in the range of one nanometer[52], enabling multi-billions of transistors per chip. At this very small scale, we cross the boundaries of classical physics to enter the realm of quantum mechanics. At this lower nanometer scale, particles behave more like waves than tennis balls. To put it very simply, it is difficult to say where exactly a wave is, and since it moves permanently, it is also hard to determine the state (1 or 0) of the wave. Consequently, if we decide to code information by such a wave and call this a "quantum bit" or "Q-Bit", then, because of its quantum nature, the Q-Bit state is the superposition of being 1 and being 0 at the same time. In short, one Q-Bit is two possibilities at the same time. If another Q-Bit is added to a quantum chip, it comes up to two Q-Bits and four possibilities at the same time! Every time a quantum bit is added to a quantum chip, its computational power doubles. The rule is exponential: if we build a quantum computer with N Q-Bits, it encompasses 2^N possibilities at the same time. While the number of Q-Bits grows linearly, the number of classical transistors required to achieve the same computational power grows exponentially. With a few hundred Q-Bits, a quantum computer could check all the atoms in the universe at the same time. A total of 300 Q-Bits are more powerful than all the computers in the world working together, a few hundred Q-Bits versus trillions of transistors! This means quantum chips can tackle so-called "exponentially difficult" problems, in an acceptable period of time. A famous example is the "RSA" (Rivest–Shamir–Adleman) method, known as the public key encryption protocol. RSA is based on the factorization of a large number into the product of two prime numbers, called the factorization problem. It has been shown that solving the RSA problem with a number coded on 1024 bits would take a few thousand years with conventional computers, while a quantum computer would do it in a few minutes. Other stunning applications are envisioned for quantum computers: more human-like AI, financial portfolio optimization, computational chemistry, drug design, logistics and scheduling, cyber-security, code breaking, circuit, software and system fault simulation, improving cancer treatment, better traffic flow, better mobile data coverage, more accurate weather forecasting, more efficient customized advertising and so on. The race has been on for more than a decade among prestigious organizations and large companies, such as MIT, IBM, Google and Intel, to build a viable quantum computer!

Box 6.1. *Quantum computing*

Sensing, exchanging, collecting, processing and communicating relevant amounts of trustable data in the BioPhyGital world are the prerequisite, as

51 Originally introduced as a rule of thumb!

52 One nanometer is one million times smaller than a human hair!

with the case for humankind, to reliable "learning and making" capabilities that we focus on next.

AI, cobotics. Standing at the bottom vertex of the AugmentChain enabler's pyramidal engine in Figure 6.5, artificial intelligence[53] and cobotics[54] (see Groove X, https://groove-x.com/en/) mimic two important capabilities of humankind, "**learning and making**", that have driven, along with symbol-based language skills, the evolution of humans for about 3 million years.

In fact, we are at the dawn of a new era: the "augmentation convergence era" that will entirely change everything that we do at an incredible pace. The way we busy ourselves to "assure our survival in ancient times" (we might say the "way we work" in more recent times), will change more in the next decades than it has in the last thousands of years, thanks to the fully-sustainable AugmentChain (see Figure 4.7). Humankind has so far experienced four ages of surviving behaviors, namely "hunting and gathering", that lasted more than a million years, "agriculture", that lasted thousands of years, "industry", that has lasted more than two centuries and "digitizing", that we began only a few decades ago, since the tail end of the 20th Century, with the advent of computers and the Internet. Moreover, clearly enough, the clock is speeding up!

The fifth, emerging, surviving behavior era is "the augmenting age" based on sustainable cross augmentation of the BioPhyGital world that will resiliently keep our planet safe, fed, healthy and powered[55], while we can generate multi-faceted and beneficial value, whether it is cognitive, emotional, spiritual, medical, environmental, financial or societal, and whether as a citizen, a family, an organization, a company or a whole country.

Obviously, being able to equip the BioPhyGital world with a "meta-brain", in other words, enabling mimesis of how the human brain

53 Luc Julia also talks about Augmented Intelligence in his book *Artificial Intelligence does not exist*.

54 Cobotics might be viewed as "collaborative robotics" between humans and robots and algorithms. Cobotics is actually a key tool of the BioPhyGital World to perform tasks altogether that neither of them could have done alone, for example the design of a special car frame; see the "Incredible invention of intuitive AI" by Maurice Conti: https://www.youtube.com/watch?v=aR5N2Jl8k14.

55 See Meersens: https://www.lafrenchfab.fr/entreprise/meersens/.

imagines, learns, creates, reasons, feels, guesses, socializes, decides and governs, either alone or as a group, is the Holy Grail on the augmentation convergence road map, already attempted with varying degrees of accomplishment or success since the very origins of AI.

Since the pioneering AI work of Turing in the 1940s, the most convincing "brain like" tasks accomplished by machines so far have been mostly limited to learning patterns based on pre-existing databases and, then recognizing these patterns in a real life situation, relying on deep (supervised) learning algorithms that mimic neural networks. For example, Amazon, Facebook and Apple are behemoths of automatic facial recognition based on pattern learning schemes, respectively for payment, social interaction and iPhone access. Emotion recognition emerged in CES 2020 as the must-have feature in connected and autonomous cars (Hyundai, Kia, Waymo, Nissan, etc.). Of course, all chatbots, Alexa, Siri and Google Home use the same deep learning analytics for speech recognition. We are, in any event, still very far from one-shot[56] learning or intuitive decision-making.

Nevertheless, checking the spectacular events most covered by the media in the last decades, we see a steady trend towards more sophistication in what machines can do, thanks to continuous increases in computing power:

– 1952 – a "kind of computer" plays tic-tac-toe (machine merely applies rules);

– 1997 – Deep Blue (IBM) beats Kasparov at chess (machine check patterns);

– 2011 – Watson (IBM) beats a human at Jeopardy (machine relies on some kind of reasoning);

– 2017 – AlphaGo by DeepMind (Google) beats the best Go players in the world (since Go has more moves than there are atoms[57] in the universe, AlphaGo has to rely on some kind of intuition to beat humans);

– 2018 – OpenAI (Microsoft) beats the world's best players at a video game, Dota 2 (machine had to make relevant strategic decisions). Microsoft invested $1 billion in 2019 in OpenAI to build artificial general intelligence

56 "One-shot learning" happens in real life when, for example, a child or a grown-up learns and memorizes something, such as a new word, a new concept and a new symbol, at once, without the need to repeat it. In 2019, machines could not do this.

57 The estimated number of atoms in the universe is 10^{80}, while a lower bound of Go number of moves is 10^{170}!

(AGI)[58] that, broadly speaking, can do almost everything the human brain can accomplish.

However, sophisticated artificial intelligence that can act like BioPhyGital world's meta-brain turns out to be useless if it is not protected against impairments and attacks, or if its governance is loose and unfair, acting as an open door to chaos. These are the issues that we tackle with the next enabling capabilities.

Cyber-security and blockchain[59] are the last key AugmentChain components of the pyramidal enabler's engine (located on the left vertex in Figure 6.5). Cyber-security and blockchain actually mimic "**protection and empowerment**" capabilities of humans who have thrived for hundreds of thousands of years, respectively referring, firstly, to the biological survival of humans and, secondly, to the social and societal sustainability of humans through appropriate governance. Cyber-security and blockchain are the BioPhyGital world's immune system and the horizontal governance/ empowerment tool (in 2020[60]) respectively. Since blockchain has already received much attention previously in this chapter (see section 6.2), we focus hereafter on cyber-security.

This book is not intended to review in detail the complex mechanisms of cryptographic algorithms, either conventional or quantum, for the protection, defense and resilience of the BioPhyGital world, nor to list categories of malware and most deadly cyber viruses; but it is intended to raise the reader's awareness that cyber-security is a job for all of us, since, as reported by Cybint Solutions[61], 95% of security breaches are the result of human behaviors[62].

As Ginni Rometty, IBM President, Chairman and CEO puts it, "Cybercrime is the greatest threat to every company and organization in the world."

58 AGI is also a target for DeepMind. Most AI experts believe that AGI will not happen for many decades, if ever.

59 See section 6.3 of this chapter.

60 As already mentioned, these technologies may evolve, but the need for protection and empowerment will remain.

61 See https://www.cybintsolutions.com/cyber-security-facts-stats/.

62 Cybercriminals infiltrate companies through the weakest link (humans), which is not the IT department.

The following facts and statistics, from Cybint Solutions, confirm that Ginni Rometty is absolutely right:

– the total cost of cybercrime in 2018 totaled $1 trillion (1,000 billion);

– there is a hacker attack every 39 seconds;

– since 2013, a total of 3,809,448 records have been stolen through breaches every day (158,727 per hour, 2,645 per minute and 44 every second);

– large-scale distributed denial of service[63] (DDoS) attacks have grown by 500% per year since 2018, due to the explosion of the IoT with 1 trillion everyday objects to be connected to the Internet by 2022;

– over 75% of the healthcare industry was infected by malware in 2019;

– 43% of cyber-attacks target small businesses;

– only 38% of global organizations claim they are prepared to handle sophisticated cyber-attacks;

– democratic processes and the world economic order are now destabilized by cyber-attacks similar to the 2016 US presidential elections and the attacks against very large companies such as Uber, Deloitte and Altran;

– despite the fact that 95% of breaches in the recent past came from only three sectors, i.e. government, retail and technology, anybody, any organization, any religious group, any gender association, any political party, and any connected thing is now a potential target of cyber-attack;

– unfilled cyber-security jobs will reach 3.5 million worldwide by 2021.

All these facts and figures mean that, with cyber vulnerability growing exponentially in scale and pace, triggered by exponential growth in the number of objects connected to the Internet and their users, cyber-attack is one of the biggest threats to every person, company, organization and country that the BioPhyGital world will face in the coming years, which could at some point close the Internet down!

63 A distributed denial-of-service (DDoS) attack is an attack in which multiple compromised computer systems attack a target, such as a server, a website or another network resource, and slows down or even shuts down the service for users of the targeted resource.

Apart from building ever more sophisticated cyber-security processes and behaviors, which are struggling to keep up with cybercriminals, another way to handle the cyber-attack threat is to better understand how cybercriminal groups such as Anonymous[64] and LulzSec[65] operate, organize, form and dig into their motives, targets and means.

In addition to genuine hacker groups, classically-organized criminal gangs are broadening their scope to the "cyber business", not only thanks to the degree of anonymity it provides but also through very efficient crime packs and services available for cyber-attack beginners, enabling them to launch high-tech attacks with very limited understanding. Virtest.com, for example, provides test beds to gauge the quality of a virus. "Black Hole Exploit" makes a business intelligence reporting dashboard available to track virus spread[66].

6.4.5. *The AugmentChain catalyst's pyramidal engine*

The catalyst's pyramidal engine shown in Figure 6.6 is fueled by levers that aim to overcome the four types of hurdle that pave the way towards the augmentation convergence of the BioPhyGital world. However, no matter how sophisticated, powerful and fast the most advanced enabling technologies described in Figure 6.5 become (AI, blockchain, cobotics, quantum computing, cyber-security, IoT AR/VR/MR, chatbots, digital twins, etc.), they will turn out to be useless, despite many billions of investment, if they are not part of an ITA value chain philosophy that dynamically overcomes the four hurdles (from the top vertex shown in Figure 4.6 and then moving clockwise): human, ethical and environmental, economic and, lastly, legal.

Google Glass is a recent example of a technology innovation in the digital age that failed to pass each of the four barriers, despite being launched by a GAFA behemoth, anything but a naive startup wanting to save the world. From the human perspective, the odd appearance of

64 Anonymous is behind cyber-attacks on the FBI, US Department of Justice and US Copyright Office, declarations of war on banks and stock exchange markets, and more recent calls to action against US President Donald Trump.
65 LulzSec stole private data from 24.6 million customers via a hack on Sony's PlayStation Network.
66 See, for example, https://www.hendersonscott.com/news/5-must-watch-cyber-security-ted-talks--66696/. For more details, see https://arxiv.org/pdf/1901.01914.pdf.

the device, the terrible UI[67] and the risk of wearing a gismo that emits carcinogenic radiation on the head and so close to the eyes for many hours a day did not live up to the hype, even among the most savvy of geeks. If likely target customers/users had been more involved in the early stage of the design, adoption and appropriation would have been better. From an ethical and environmental vantage point, power consumption combined with a potentially invasive gadget that intrudes into your personal life (and other people's) and peers seamlessly, sometimes secretly, was definitely a no-go! Economically, Google Glass had neither a convincing function nor a clear expected outcome (see Figure 6.4 and the related comments). The key to creating a disruptive product is to try to meet a perceived demand or solve a known problem/pain point, not build it first and then try to find a market. On top of a real lack of any business target, the prohibitive price of $1,500 per unit stopped sales from taking off. Finally, from a legal standpoint, such a massive personal data hoover was doomed to inherently violate the General Data Protection Regulation (GDPR)[68].

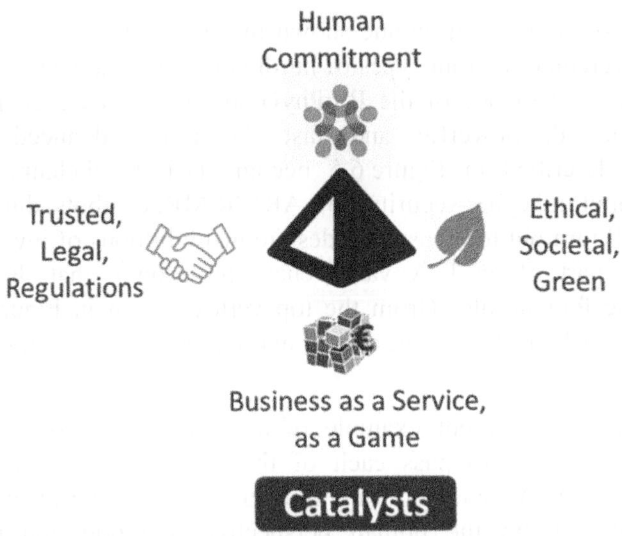

Figure 6.6. *The AugmentChain catalyst's pyramidal engine*

67 User interface.

68 https://gdpr-info.eu/; GDPR (General Data Protection Regulation) is a regulation in EU law on data protection and privacy for all individual citizens of the European Union (EU) and the European Economic Area (EEA).

Human commitment/engagement. Human catalyst is paramount. It is the central factor.

What is under the hood of getting motivated people to engage in different types of activities, such as buying, selling, working, making and innovating, at the time of the six great transitions (digital, industrial, environmental, societal, medical and quantum) causing upheaval?

It is not the scope of this book to go deeper into human motivation, like in the work of David C. McClelland[69], but to merely cast some light on the human component of ITA value chain sustainability.

According to neuroscientists, humans are hardwired to:

– be connected;

– seek pleasure;

– avoid pain;

– be part of something bigger than who they are;

– be "heroes" of big causes (i.e. to give special sense to what we do, beyond just making profit out of our deeds).

Stimulating each or a combination of these feelings may most likely, and to some extent, foster engagement/involvement of users, patients, employees and citizens, regardless of age, gender and culture towards multi-faceted ITA value creation/enrichment.

Collaborating and/or sharing via a social network (Twitter, Facebook, Instagram, WeChat), carpooling (BlaBlaCar), meal sharing[70], etc., are nothing but ways to "connect" and subsequently to engage.

Reluctance to change, for example, or first emphasizing all the underlying risks of an opportunity instead of prioritizing what you can gain

69 https://books.google.fr/books/about/Human_Motivation.html?id=vic4AAAAIAAJ&redir_esc=y; "Scientific understanding of motives and their interaction, David C. McClelland argues, contributes to understanding of such diverse and important phenomena as the rise and fall of civilizations, the underlying causes of war, the rate of economic development, the nature of leadership, the reasons for authoritarian or democratic governing styles, the determinants of success in management and the factors responsible for health and illness."

70 https://www.mealsharing.com/.

from a new product, service and user experience, is a trivial way to dodge pain and can trigger engagement, for the least optimistic among us!

Gamification can make you feel that you belong to something bigger than yourself and boost engagement. Pain Squad[71], for example, is a mobile game designed to help children better fight cancer by providing the patients with a "purpose" that helps them to transcend their sickness and engage with the therapy or treatment.

The more and the earlier you get involved in the creation, deployment and life of a product, a service and more generally a user experience, the more you will be engaged, especially when the involvement is genuine and user requirements/feedback, as well as opinions, are genuinely heard and acted upon via continuous improvement of a hyper-personalized product/service/UX/UI design.

Involvement is also triggered by empowerment: making sure that people see the benefits and are incentivized in a proper way. What is the value proposal for me at the end of the day? How and how much am I encouraged to even start to use a "pay as you live" multi-risk insurance policy[72]? Your insurance benefits and rates change according to the way you live. To return to the pleasure/pain motivation duality, one potential pain will be to appear as you truly are. Along with the human factor, ethics and ecology play key roles in the sustainability of ITA value chains.

Ethical, societal, green levers, shown on the right vertex of the catalyst's pyramid in Figure 6.6, aim to remove environmental and moral barriers to a resilient and perennial AugmentChain!

On the ethical side, actions have been undertaken by social networks, in particular Facebook, to mitigate the "shame for fame" or "fake for fame" behaviors that can severely disrupt the day-to-day life of potential scapegoats.

On the ecological/environmental side, it is obvious, for example, that despite the potential social, political, ethical and behavioral empowerment

---

71 https://fr.slideshare.net/manumelwin/pain-squad-social-gamification-manu-melwin-joy.

72 See, for example, https://www.realinsurance.com.au/car-insurance/pay-as-you-drive.

that a blockchain may provide citizens with, as the "trust machine", its wide adoption and durability are just beginning. At a time when "climate change is killing more and more people"[73], drastically curbing greenhouse gas emissions – from any source (livestock, industry, cars, IT, people) – has become a matter of survival.

Clearing economic obstacles is also a *sine qua non* condition to creating value.

Business as a Service and Business as a Game. The bottom vertex in Figure 6.6 shows two levers (Business as a Service and Business as a Game), which aim to overcome major economic obstacles in the path towards sustainable AugmentChain.

ITA value chains are driven by the so-called "API economy" that turns a business or organization into a platform that delivers almost anything as a service[74] (XaaS).

An API (Application Programming Interface) is designed to create a new user experience, in order to authenticate, negotiate, connect people, places and devices, share data and use third-party analytics.

For example, if we look at Apple, the total number of iOS APIs working on one of the three devices, namely iPhone, iPad or iPod touch, has exploded to 2,200,000 in 2019, starting from 800 in 2008! This is a multiplication by a factor of 2,750 over a span of 11 years.

Platforms such as Amazon, Uber and Netflix drastically amplify value creation because they foster "business as a service" (BaaS) ecosystems inside and outside of the enterprise. They facilitate the creation and/or exchange of goods, services and social connectivity so that everybody benefits from it.

Business as a Service turns out to be a combination of APIs, platforms and the delivery of user experience as a service. It has penetrated all facets of

73 *The Economist*, August 2, 2019.

74 XaaS refers to the delivery of "anything as a service". It encompasses products, tools and technologies that vendors now deliver to users as a service over a network – typically the Internet – rather than providing locally or on-site within an enterprise.

life, with Mobility_aaS[75], Voice_aaS (Alexa, Google Home, Siri), Health_aaS (Rush Health[76]), Bank_aaS[77], Commerce_aaS (JD.com, Amazon, Walmart), etc. The generic business model of AugmentChain is ITA_aaS (Intangible Asset as a Service).

BaaS is thus a genuine catalyst of the ITA value chain take-off and growth. Gamification is another important example.

We emphasize hereafter key aspects of how gamification can boost ITA business and monetization, based on the enlightening work of Yukai Chu[78].

As Yukai Chu puts it, "Gamification is a concept that involves picking the best part from games and applying them in our usual activities, in particular business. Gamification is human-focused design, as opposed to function-focused design".

Yukai Chu considers that gamification DNA is based on several core drivers that can improve our everyday life. Here are some of the core drivers that appear to be the most relevant to catalyzing ITA businesses:

– "Epic meaning and calling" (Being a player in an ITA value chain will give you the feeling that you are part of something bigger than yourself; Zero Mass Water[79] has built the first hydro panel that converts sunlight and air into drinking water. When you use it, you truly feel that you have contributed to saving the planet's resources.)

– "Development and accomplishment" (AugmentChain will help you improve by any means. For example, the value proposal of the Nike Plus SportWatch is truly focusing and permanently checking and improving your own workouts[80].)

– "Social influence and relatedness" (AugmentChain will influence what you do and buy based on what other people think, do and say. Opower[81], for

75 "The key point about MaaS is that it envisages users buying transport services as packages based on their needs instead of buying the means of transport", https://www.sciencedirect.com/science/article/pii/S2352146516302836. See also https://www.gov.uk/government/publications/future-of-mobility-mobility-as-a-service.

76 https://www.rush-health.com/RHAWeb/default.aspx.

77 http://www.bank-as-a-service.com/.

78 https://learn.g2.com/gamification.

79 https://www.zeromasswater.com/na/.

80 See https://www.cnet.com/reviews/nike-plus-sportwatch-gps-review/.

81 https://ux.opower.com/#.

example, helps people lower their utility bills by showing them how their neighbors are doing it.)

In addition to economic barriers, the final hurdles to clear – and their nature makes this non-negotiable – are regulation and law.

Trusted, legal regulations. At a time when privacy violations may become a real threat for citizens (facial recognition, intrusive and personal data-driven APIs, etc.), the levers shown on the left vertex in Figure 6.6, which aim to restore trust in many different ways – potentially including regulation and legal means – are paramount.

Trust will be a key competitive differentiator in the augmentation era.

A coercive way to increase trust and lay some legal foundations for ITA value chain sustainability is to force companies and organizations to abide by regulations and laws that are enforced to avoid opacity, privacy violations, unfair social interactions and questionable business behaviors.

You will be more inclined to share your data with companies, platforms, social networks and organizations if there are laws and regulations that aim to equalize the three asymmetries (see section 6.2) of trust, information and revenues. An article in the August 2, 2019 edition of *The International New York Times*, entitled "In the Haze of Privacy Violations, FaceApp is the future", perfectly describes what we need to prevent. FaceApp provides age-advanced portraits based on genuine original younger pictures of you, and makes them available to anyone, for free. The catch is that users must agree to grant Facebook a free license to their pictures that is "perpetual, irrevocable, non-exclusive, royalty-free and worldwide". Is this right? When released in the summer of 2019, FaceApp went so viral that users did not pay so much attention to its GCU (General Conditions of Use), which are obviously questionable from a privacy standpoint, and also when viewed from the unfair benefit of shared ITA (your photos) for the social network. As pointed out by *The International New York Times*, the "cartoonish" style of you getting older combined with some mysterious power of AI that behaves like an oracle, bringing back news from the future (i.e. who you become), explains, to some extent, why users quickly and addictively fell for FaceApp, giving it their own blessing to share their privacy.

The time has come for a pivotal move, requested by citizens, towards a sustainable ITA value chain by giving control over data to data sources.

This is where laws and regulations come into the picture, with an extra subtle mission, on top of setting legal barriers against privacy intruders. It helps users to mitigate their obsessive addiction to screens and virtual life. The GDPR is the most important regulation, coming into effect on May 25, 2018, to protect privacy for all individual citizens of the European Union (EU) and the European Economic Area (EEA).

It falls outside the scope of this book to detail the 173 recitals of the GDPR, but let us emphasize some of its key aspects[82]:

GDPR driving principles[83] raise shields against privacy thieves:

– data processing requires a legal basis (lawfulness);

– data subjects should be informed (transparency);

– data should be used only for specific purposes (purpose limitations);

– only process the necessary amounts of data to achieve the claimed purpose (data minimization);

– delete data that is no longer required (storage minimization);

– the controller is responsible for demonstrating compliance (accountability principle).

Failure to comply with the GDPR triggers fines, at the lower level, of up to €10 million or 2% of the worldwide annual revenue in the previous financial year, whichever is higher. At the higher level, the fine is up to €20 million, or 4% of the worldwide annual revenue in the previous financial year, whichever is higher[84].

Nevertheless, implementing the GDPR will be seized as a unique opportunity for organizations to acquire the rights to use, process, enrich and share data, for the sake of business value and the sustainability of the ITA value chain.

---

82 For more details, see https://gdpr-info.eu/.

83 Excerpts from Bird & Bird Law firm reports, https://www.twobirds.com/.

84 See https://www.gdpreu.org/compliance/fines-and-penalties/.

6.5. Conclusion and takeaways

As the rich journey through the "AugmentChain" comes to its end, we summarize some key takeaways, based on the "big picture" given in Figure 6.7, which combines all the bricks and underlying principles explained in this chapter.

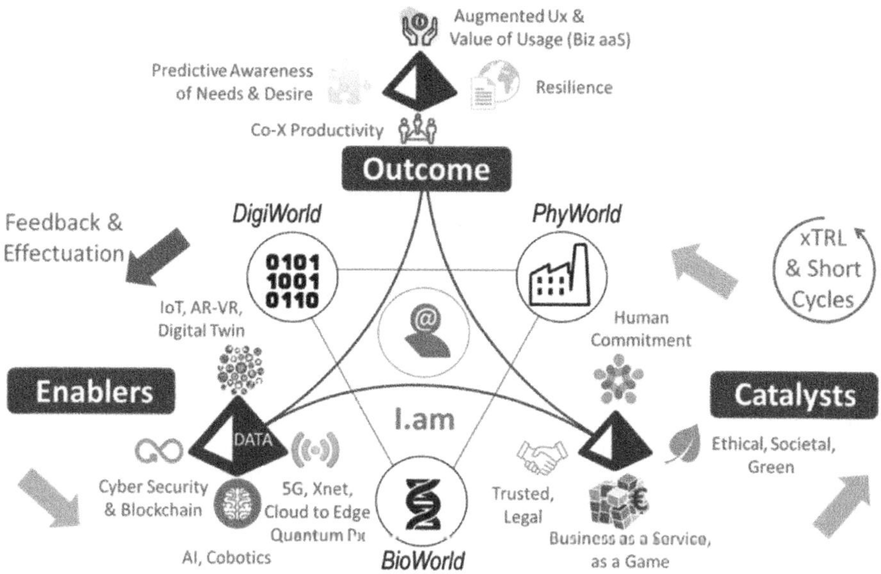

Figure 6.7. *The sustainable ITA augmentation value chain, called the "AugmentChain". For a color version of this figure, see www.iste.co.uk/duvaut/IAM.zip*

We are at the dawn of a new era: the "augmentation convergence era" that will entirely change everything that we do[85] at an incredible pace. The way we make ourselves busy to "assure our survival in ancient times", or the "way we work" as we describe it nowadays, will change more in the next few decades than in the last several thousand years, thanks to the AugmentChain.

It has the potential to unleash the multi-faceted value of intangible assets (ITA) – financial, societal, cognitive, economic, medical, industrial and environmental – in a sustainable manner for people, society, companies, organizations and the planet.

85 The augmented operating theater is a concrete example of the AugmentChain applied to surgery innovation, https://www.imt.fr/en/company-offers/surgical-cockpit-innovation-chair/.

Thanks to powerful engines (digital today, quantum tomorrow), we are engaged in a co-evolution/convergence of a "BioPhyGital" world, permanently connected, striving to deliver augmented UX in the form of "existential intangible assets", such as your next need and wish. This is the "Internet of Augmented Me, I.AM".

The augmentation convergence of the BioPhyGital world is driven by three "pyramidal" engines (outcomes, enablers and catalysts) that work together in short cycles.

Augmentation convergence is going through enrichment loops of "effectuation" that take advantage of feedback from each of the three worlds (Bio, Phy and Digital), in order to permanently improve the efficiency and the resilience of the user experience delivered.

Motivated by the quest for a new sustainable digital era where major platform asymmetries (lack of trust, information and benefits, as sometimes encountered in GAFA, NATU and BATX) could be balanced out, the AugmentChain outcomes engine is more driven by sustainability, sense, epic meaning, accomplishments, well-being, usage, human and citizens' rights than trivial value-driven targets such as cost reduction, increased revenues, better efficiencies or new business models.

Each of the four vertices of the enabler's pyramid mimics key human capacities that have triggered and fostered the evolution of humankind for hundreds of thousands of years: "sensing and interacting", "communicating and processing", "learning and making" and "protecting and empowering".

However, while the most advanced enabling technologies may be sophisticated, powerful and fast, they are useless, despite many billions in investment, if they are not paired with catalysts, aimed at overcoming the four types of hurdle in the path towards large scale and lasting adoption, namely human, ethical–environmental, economic and legal hurdles.

6.6. List of acronyms

aaS: as a Service

AGI: Artificial General Intelligence

AI: Artificial Intelligence

API: Application Programming Interface

AR: Augmented Reality

BATX: Baidu, Alibaba, Tencent, Xiaomi

DDoS: Distributed Denial of Service

EEA: European Economic Area

eMBB: enhanced Mobile Broadband

EU: European Union

GAFA: Google, Amazon, Facebook, Apple

GCU: General Conditions of Use

GDPR: General Data Protection Regulation

I2V: Invisible to Visible

I.AM: Internet of Augmented Me

IoT: Internet of Things

ITA: Intangible Assets

ITU: International Telecom Union

KPI: Key Performance Indicators

mMTC: massive Machine-Type Communications

MR: Mixed Reality

NATU: Netflix, Airbnb, Tesla, Uber

QBit: Quantum Bit

URLLC: Ultra-Reliable Low-Latency Communications

V2V: Vehicle to Vehicle

V2X: Vehicle to Anything

VR: Virtual Reality

VUI: Voice User Interface

References

Books

[ALE 77] ALEXANDER C., *A Pattern Language: Towns, Buildings, Construction*, Oxford University Press, London, 1977.

[BEN 97] BENYUS J., *Biomimicry: Innovations Inspirad by Nature*, HyperCollins Publishers Inc., New York, 1997.

[DOS 09] DOSTALER G., *Keynes, par delà l'économie,* Thierry Magnier Editions, Paris, 2009.

[GOR 10] GORDON B., GATES B., *Your Life Uploaded*, Plume, 2010.

[PIT 18] PITRON G., *La guerre des métaux rares : la face cachée de la transition énergétique*, Editions Les Liens qui Libèrent, Paris, 2018.

[RAJ 19] RAJKUMAR B., SATISH NARAYANA S., *Fog and Edge Computing – Principles and Paradigms*, Wiley, 2019.

[SCH 16] SCHWAB K., *The Fourth Industrial Revolution*, World Economic Forum, 2016.

[SHA 18] SHAPIRO G., *Ninja Future: Secrets to Success in New World of Innovation*, William Morrow, 2018.

[STI 10] STIGLITZ E.J., *Freefall*, W.W. Norton & Company, 2010.

Conferences, magazine articles

[DEA 18] DE ASSUNCAO M.D., DA SILVA VEITH A., RAJKUMAR B., "Distributed data stream processing and edge computing: A survey on resource elasticity and future directions", *Journal of Network and Computer Applications*, Elsevier, vol. 103, pp. 1–17, 2018.

[LOP 15] LOPEZ P., DATTA A, BARCELLOS M. *et al.*, "Edge-centric computing: Vision and challenges", *Computer Communication Review*, vol. 45, no. 5, 2015.

[STR 01] STREITZ N., "Mental vs physical disappearance, the challenge of interacting with disappearing computers", *ACM CHI Conference on Human Factors in Computing Systems*, Seattle, USA, March 31–April 5, 2001.

[WEI 91] WEISER M., "The computer for the 21th Century", *Scientific American,* September 1991.

Unpublished Ph.Ds & Reports, etc.

[CHR 18] CHRIQUI V., HABABOU H., Booking token unit protocol, White paper, March 2018.

[DOL 17] DOLUI K., DATTA S.K., "Comparison of edge computing implementations: Fog computing, cloudlet and mobile edge computing", *IEEE*, 2017.

[PRE 18] PREMSANKAR G., DI FRANCESCO M., TALEB T., "Edge computing for the internet of things: A case study", *IEEE*, 2018.

[SHI 16] SHI W., CAO J., ZHANG Q. *et al.*, "Edge computing: Vision and challenges", *IEEE*, 2016.

Web links

[AMB 15] AMBERBER E., 'Banking is necessary, banks are not' – 7 quotes from Bill Gates on mobile banking. Available at: https://yourstory.com/2015/01/quotes-bill-gates-mobile-banking?utm_pageloadtype=scroll, 2015.

[NAK 18] NAKAMOTO S., Bitcoin: A peer-to-peer electronic cash system. Available at: https://bitcoin.org/bitcoin.pdf, 2018.

[WIL 12] WILLETT J.R., The Second Bitcoin Whitepaper. Available at: https://assets.ctfassets.net/sdlntm3tthp6/resource-asset-r362/eca527b3c2072972b0d34b278dca2a33/0b74d5e3-1199-443a-b1e6-7fabe0dac808.pdf, 2012.

[ZER 17] ZERANSKI T., Energy companies join forces with Rocky Mountain Institute and Grid Singularity to launch global blockchain initiative for energy. Available at: https://rmi.org/press-release/press-release-energy-web-foundation-launch/, 2017.

Index

D

E, F

G, H, I

T, U, V

W, X, Z

Other titles from

iSTE

in

Computer Engineering

2020

LAFFLY Dominique
TORUS 1 – Toward an Open Resource Using Services: Cloud Computing for Environmental Data
TORUS 2 - Toward an Open Resource Using Services: Cloud Computing for Environmental Data
TORUS 3 - Toward an Open Resource Using Services: Cloud Computing for Environmental Data

LAURENT Anne, LAURENT Dominique, MADERA Cédrine
Data Lakes
(Databases and Big Data Set – Volume 2)

OULHADJ Hamouche, DAACHI Boubaker, MENASRI Riad
Metaheuristics for Robotics
(Optimization Heuristics Set – Volume 2)

SADIQUI Ali
Computer Network Security

2019

BESBES Walid, DHOUIB Diala, WASSAN Niaz, MARREKCHI Emna
Solving Transport Problems: Towards Green Logistics

CLERC Maurice
Iterative Optimizers: Difficulty Measures and Benchmarks

GHLALA Riadh
Analytic SQL in SQL Server 2014/2016

TOUNSI Wiem
Cyber-Vigilance and Digital Trust: Cyber Security in the Era of Cloud Computing and IoT

2018

ANDRO Mathieu
Digital Libraries and Crowdsourcing
(Digital Tools and Uses Set – Volume 5)

ARNALDI Bruno, GUITTON Pascal, MOREAU Guillaume
Virtual Reality and Augmented Reality: Myths and Realities

BERTHIER Thierry, TEBOUL Bruno
From Digital Traces to Algorithmic Projections

CARDON Alain
Beyond Artificial Intelligence: From Human Consciousness to Artificial Consciousness

HOMAYOUNI S. Mahdi, FONTES Dalila B.M.M.
Metaheuristics for Maritime Operations
(Optimization Heuristics Set – Volume 1)

JEANSOULIN Robert
JavaScript and Open Data

PIVERT Olivier
NoSQL Data Models: Trends and Challenges
(Databases and Big Data Set – Volume 1)

SEDKAOUI Soraya
Data Analytics and Big Data

SALEH Imad, AMMI Mehdi, SZONIECKY Samuel
Challenges of the Internet of Things: Technology, Use, Ethics
(Digital Tools and Uses Set – Volume 7)

SZONIECKY Samuel
Ecosystems Knowledge: Modeling and Analysis Method for Information and Communication
(Digital Tools and Uses Set – Volume 6)

2017

BENMAMMAR Badr
Concurrent, Real-Time and Distributed Programming in Java

HÉLIODORE Frédéric, NAKIB Amir, ISMAIL Boussaad, OUCHRAA Salma, SCHMITT Laurent
Metaheuristics for Intelligent Electrical Networks
(Metaheuristics Set – Volume 10)

MA Haiping, SIMON Dan
Evolutionary Computation with Biogeography-based Optimization
(Metaheuristics Set – Volume 8)

PÉTROWSKI Alain, BEN-HAMIDA Sana
Evolutionary Algorithms
(Metaheuristics Set – Volume 9)

PAI G A Vijayalakshmi
Metaheuristics for Portfolio Optimization
(Metaheuristics Set – Volume 11)

2016

BLUM Christian, FESTA Paola
Metaheuristics for String Problems in Bio-informatics
(Metaheuristics Set – Volume 6)

DEROUSSI Laurent
Metaheuristics for Logistics
(Metaheuristics Set – Volume 4)

DHAENENS Clarisse and JOURDAN Laetitia
Metaheuristics for Big Data
(Metaheuristics Set – Volume 5)

LABADIE Nacima, PRINS Christian, PRODHON Caroline
Metaheuristics for Vehicle Routing Problems
(Metaheuristics Set – Volume 3)

LEROY Laure
Eyestrain Reduction in Stereoscopy

LUTTON Evelyne, PERROT Nathalie, TONDA Albert
Evolutionary Algorithms for Food Science and Technology
(Metaheuristics Set – Volume 7)

MAGOULÈS Frédéric, ZHAO Hai-Xiang
Data Mining and Machine Learning in Building Energy Analysis

RIGO Michel
Advanced Graph Theory and Combinatorics

2015

BARBIER Franck, RECOUSSINE Jean-Luc
COBOL Software Modernization: From Principles to Implementation with the BLU AGE® Method

CHEN Ken
Performance Evaluation by Simulation and Analysis with Applications to Computer Networks

CLERC Maurice
Guided Randomness in Optimization
(Metaheuristics Set – Volume 1)

DURAND Nicolas, GIANAZZA David, GOTTELAND Jean-Baptiste, ALLIOT Jean-Marc
Metaheuristics for Air Traffic Management
(Metaheuristics Set – Volume 2)

MAGOULÈS Frédéric, ROUX François-Xavier, HOUZEAUX Guillaume
Parallel Scientific Computing

MUNEESAWANG Paisarn, YAMMEN Suchart
Visual Inspection Technology in the Hard Disk Drive Industry

2014

BOULANGER Jean-Louis
Formal Methods Applied to Industrial Complex Systems

BOULANGER Jean-Louis
Formal Methods Applied to Complex Systems:Implementation of the B Method

GARDI Frédéric, BENOIST Thierry, DARLAY Julien, ESTELLON Bertrand, MEGEL Romain
Mathematical Programming Solver based on Local Search

KRICHEN Saoussen, CHAOUACHI Jouhaina
Graph-related Optimization and Decision Support Systems

LARRIEU Nicolas, VARET Antoine
Rapid Prototyping of Software for Avionics Systems: Model-oriented Approaches for Complex Systems Certification

OUSSALAH Mourad Chabane
Software Architecture 1
Software Architecture 2

PASCHOS Vangelis Th
Combinatorial Optimization – 3-volume series, 2nd Edition
Concepts of Combinatorial Optimization – Volume 1, 2nd Edition
Problems and New Approaches – Volume 2, 2nd Edition
Applications of Combinatorial Optimization – Volume 3, 2nd Edition

QUESNEL Flavien
Scheduling of Large-scale Virtualized Infrastructures: Toward Cooperative Management

RIGO Michel
Formal Languages, Automata and Numeration Systems 1: Introduction to Combinatorics on Words
Formal Languages, Automata and Numeration Systems 2: Applications to Recognizability and Decidability

SAINT-DIZIER Patrick
Musical Rhetoric: Foundations and Annotation Schemes

TOUATI Sid, DE DINECHIN Benoit
Advanced Backend Optimization

2013

ANDRÉ Etienne, SOULAT Romain
The Inverse Method: Parametric Verification of Real-time Embedded Systems

BOULANGER Jean-Louis
Safety Management for Software-based Equipment

DELAHAYE Daniel, PUECHMOREL Stéphane
Modeling and Optimization of Air Traffic

FRANCOPOULO Gil
LMF — Lexical Markup Framework

GHÉDIRA Khaled
Constraint Satisfaction Problems

ROCHANGE Christine, UHRIG Sascha, SAINRAT Pascal
Time-Predictable Architectures

WAHBI Mohamed
Algorithms and Ordering Heuristics for Distributed Constraint Satisfaction Problems

ZELM Martin *et al.*
Enterprise Interoperability

2012

ARBOLEDA Hugo, ROYER Jean-Claude
Model-Driven and Software Product Line Engineering

BLANCHET Gérard, DUPOUY Bertrand
Computer Architecture

BOULANGER Jean-Louis
Industrial Use of Formal Methods: Formal Verification

BOULANGER Jean-Louis
Formal Method: Industrial Use from Model to the Code

CALVARY Gaëlle, DELOT Thierry, SÈDES Florence, TIGLI Jean-Yves
Computer Science and Ambient Intelligence

MAHOUT Vincent
Assembly Language Programming: ARM Cortex-M3 2.0: Organization, Innovation and Territory

MARLET Renaud
Program Specialization

SOTO Maria, SEVAUX Marc, ROSSI André, LAURENT Johann
Memory Allocation Problems in Embedded Systems: Optimization Methods

2011

BICHOT Charles-Edmond, SIARRY Patrick
Graph Partitioning

BOULANGER Jean-Louis
Static Analysis of Software: The Abstract Interpretation

CAFERRA Ricardo
Logic for Computer Science and Artificial Intelligence

HOMES Bernard
Fundamentals of Software Testing

KORDON Fabrice, HADDAD Serge, PAUTET Laurent, PETRUCCI Laure
Distributed Systems: Design and Algorithms

KORDON Fabrice, HADDAD Serge, PAUTET Laurent, PETRUCCI Laure
Models and Analysis in Distributed Systems

LORCA Xavier
Tree-based Graph Partitioning Constraint

TRUCHET Charlotte, ASSAYAG Gerard
Constraint Programming in Music

VICAT-BLANC PRIMET Pascale *et al.*
Computing Networks: From Cluster to Cloud Computing

2010

AUDIBERT Pierre
Mathematics for Informatics and Computer Science

BABAU Jean-Philippe *et al.*
Model Driven Engineering for Distributed Real-Time Embedded Systems

BOULANGER Jean-Louis
Safety of Computer Architectures

MONMARCHE Nicolas *et al.*
Artificial Ants

PANETTO Hervé, BOUDJLIDA Nacer
Interoperability for Enterprise Software and Applications 2010

SIGAUD Olivier *et al.*
Markov Decision Processes in Artificial Intelligence

SOLNON Christine
Ant Colony Optimization and Constraint Programming

AUBRUN Christophe, SIMON Daniel, SONG Ye-Qiong *et al.*
Co-design Approaches for Dependable Networked Control Systems

2009

FOURNIER Jean-Claude
Graph Theory and Applications

GUEDON Jeanpierre
The Mojette Transform / Theory and Applications

JARD Claude, ROUX Olivier
Communicating Embedded Systems / Software and Design

LECOUTRE Christophe
Constraint Networks / Targeting Simplicity for Techniques and Algorithms

2008

BANÂTRE Michel, MARRÓN Pedro José, OLLERO Hannibal, WOLITZ Adam
Cooperating Embedded Systems and Wireless Sensor Networks

MERZ Stephan, NAVET Nicolas
Modeling and Verification of Real-time Systems

PASCHOS Vangelis Th
Combinatorial Optimization and Theoretical Computer Science: Interfaces and Perspectives

WALDNER Jean-Baptiste
Nanocomputers and Swarm Intelligence

2007

BENHAMOU Frédéric, JUSSIEN Narendra, O'SULLIVAN Barry
Trends in Constraint Programming

JUSSIEN Narendra
A TO Z OF SUDOKU

2006

BABAU Jean-Philippe *et al.*
From MDD Concepts to Experiments and Illustrations – DRES 2006

HABRIAS Henri, FRAPPIER Marc
Software Specification Methods

MURAT Cecile, PASCHOS Vangelis Th
Probabilistic Combinatorial Optimization on Graphs

PANETTO Hervé, BOUDJLIDA Nacer
Interoperability for Enterprise Software and Applications 2006 / IFAC-IFIP I-ESA'2006

2005

GÉRARD Sébastien *et al.*
Model Driven Engineering for Distributed Real Time Embedded Systems

PANETTO Hervé
Interoperability of Enterprise Software and Applications 2005

Printed and bound by CPI Group (UK) Ltd, Croydon, CR0 4YY

13/09/2026

14961035-0004